# Military Victory Beyond the Battlefield

# VOICES IN INTERNATIONAL RELATIONS

*Voices in International Relations*, published under the auspices of the European International Studies Association (EISA), furthers the development of research at the frontiers of International Relations (IR). It expands the remit of the field by including innovative scholarship that broadens key debates in the discipline, but it is more interested in reconfiguring such debates by approaching them from inside and outside the conventional core. Thematically, we aim to publish research that pushes the limits of IR conventionally defined from within and connects it to debates developing outside the discipline. We are committed to furthering diversity and inclusion in terms of authorship, location, topics, and approaches from both inside and outside Europe. We have an inclusive approach to neighbouring disciplines, be it sociology, history, anthropology, geography, economics, political theory, or law.

For more information about the EISA visit EISA (eisa-net.org)



***Titles in the series***

*Security in Crisis*
*Planetary Emergence and the Technopolitics of Crisis Management*
Columba Peoples
*Military Victory Beyond the Battlefield*
*Outside Wartime*
Mirko Palestrino

# Military Victory Beyond the Battlefield

## *Outside Wartime*

Mirko Palestrino

EISA

OXFORD
UNIVERSITY PRESS

OXFORD
UNIVERSITY PRESS

Great Clarendon Street, Oxford, OX2 6DP,
United Kingdom

Oxford University Press is a department of the University of Oxford.
It furthers the University's objective of excellence in research, scholarship,
and education by publishing worldwide. Oxford is a registered trade mark of
Oxford University Press in the UK and in certain other countries.



Published in the United States of America by Oxford University Press
198 Madison Avenue, New York, NY 10016, United States of America

British Library Cataloguing in Publication Data
Data available

Library of Congress Control Number: 2025935501

ISBN 9780198955788

DOI: 10.1093/9780198955818.001.0001

The manufacturer's authorised representative in the EU for product safety is
Oxford University Press España S.A., Parque Empresarial San Fernando de Henares,
Avenida de Castilla, 2—28830 Madrid (www.oup.es/en or product.safety@oup.com).
OUP España S.A. also acts as importer into Spain of products made by the manufacturer.

*A Mamma e Papà*

*Grazie*

# Acknowledgements

As ironic as it might sound for a book that goes out of its way to challenge the notions of temporal boundaries, beginnings, and endings, writing these acknowledgements does feel like the end of a more than five-year-long journey. Along the way, I sometimes felt disoriented, often exhausted, but never alone. Thanks to each of the many people whose constant help, support, encouragement, guidance, and company I could count on. This book is better because of you.

Special thanks to Kimberly Hutchings and Nivi Manchanda for their wise supervision, their endless patience, and their unwavering support during my doctoral studies and research, which resulted in this manuscript. I am lucky enough to get to work with both of them now and to keep learning from two of the most inspiring and generous scholars and colleagues I have ever met. Thank you also to Andrew R. Hom and Jef Huysmans, who examined the thesis version of this work with kindness and care, shaped my thinking and scholarship, and provided invaluable comments and feedback during and after my viva. Jef has continued to mentor me since then and I cannot thank him enough for all his guidance and support. I owe special thanks also to Tarak Barkawi and Cian O'Driscoll, whose work inspired mine since its inception, and who offered exceptional feedback and advice multiple times throughout my academic career out of their goodwill.

I couldn't have pursued a doctorate, nor written this book if not for the generous funding I received as a student—the Ermenegildo Zegna Founder's Scholarship, which supported my research Master's at LSE, and the departmental scholarship of the School of Politics and International Relations (SPIR) at QMUL, which funded my PhD. SPIR and QMUL also provided additional funding to attend the numerous conferences where I discussed drafts of individual chapters and received instrumental feedback. Thanks also to the incredible staff at the Liddle Hart Centre for Military Archives, King's College London for helping me navigate the intricate world of UK doctrinal architecture.

Speaking of the type of help without which this manuscript would not have seen the light, I am forever grateful and indebted to Italo Brandimarte and Matthew Barnfield. Academics often joke about silent and secret co-authors, and Italo is most certainly mine. Italo: thank you for talking through ideas, patiently reading and improving every single piece of work I write, and sharing all the joys and pains of academic life—including being mistaken as the same person (or each other's 'other Italian guy') from the LSE days onwards. Most importantly, thank you for being my best friend. Matthew is the most brilliant political scientist I know and, even more impressively, one of the few who enjoys theory as much as I do. Matt: thank you for reading, and re-reading (and re-reading), multiple drafts of every single chapter, copyediting the entire manuscript, and helping me fix my overcomplicated syntax, abuse of passive forms, and obscure jargon. Thank you also for all the flat whites, vegan restaurant-scouting, and for being such a good friend.

This book has also benefited from exchanges, feedback, and comments of many other academics and friends. Shruti Balaji, Anna Finiguerra, Benedetta Zocchi, Jaakko Heiskanen, Lucy Kneebone, Meena Masood, Paul Kirby, and Tarsis Brito commented and read over drafts of individual chapters. I owe special thanks to Anna, with whom I shared not only ideas, drafts, and job applications, but also a flat for three years—if London feels like home now it is also thanks to you! Joanne Yao and Rick Saull oversaw two PhD progression reviews and offered invaluable comments to multiple chapters as well as career advice. I am also grateful for comments from Alex Blanchard, Alex Gould, Alex Stoffel, Alice Engelhard, Anna Leander, Angus McNelly, Chris Rossdale, Dan Öberg, Elke Schwarz, James Eastwood, Julia Costa López, Jean-François Drolet, Kate Hall, Kenneth MacLeish, Lasse Thomassen, Lauren Wilcox, Matthew Ford, Nina Krickel-Choi, Rahul Rao, Stefano Guzzini, Tasniem Anwar, Timor Landherr, Victoria Basham, as well as participants of various editions of the DoingIPS Seminar Series in London, the DoingIPS Early Career Workshops, the TheoryLab Seminar Series (QMUL), and two ad hoc workshops centred around specific chapters at the Universidad Nacional de Educación a Distancia (UNED) in Madrid, and the European University Institute (EUI) in Florence. For these workshops I also want to thank Alejandro Fernández-Roldán Díaz and Stefano Guzzini for inviting and hosting me at UNED and EUI respectively.

I am very thankful also to Oxford University Press, especially the editorial team of *Voices in International Relations*—Debbie Lisle, Tanja Aalberts, Anna Leander, and Laura Sjoberg—who believed in this project and generously supported it. I am especially grateful to Debbie, who went above and beyond her editorial duties to help me make this book better and navigate the publication process. Thank you also to Dominic Byatt and Phoebe Aldridge-Turner for overseeing this manuscript's journey all the way through production, and to the two anonymous referees for being exceptional peer reviewers.

I started and finished writing this book in SPIR at Queen Mary, first as a PhD student and then as a Lecturer. I could have not asked for a better academic home, and I am enormously grateful to everyone in the School for being so supportive and collegial. I owe special thanks to Elke Schwarz, Joe Hoover, and Martin Coward for their mentorship, and to Adam Elliot-Cooper, Aurora Ganz, Diego de Merich, Giulia Carabelli, Jaakko Heiskanen, Layli Uddin, Niharika Pandit, Rachel Humphris, Raffaele Bazurli, Rowan Lubbock, Sharri Plonski, and Sophie Harman for their help, support, and friendship. Thanks also to my fellow PhD friends with whom I shared my doctoral journey, in particular Alex Stoffel, Benedetta Zocchi, Farai Chipato, Fawzia Gibson-Fall, Félix Hernández Fernández, Hannah Owens, Ida Roland Birkvad, Janina Pescinski, Karin Narita, Lucy Kneebone, Marija Antanavičiūtė, Meena Masood, Salomé Ietter, and Timor Landherr. A shout out to Marija, Matt, Janina, Salomé, and Tarsis for also sharing several runs and to Anna, Hannah, Meena, and Shruti for the coffees, meals, and various mischiefs. And of course, to Caio, for the Cambridge adventures and the sausage dog videos!

Finally, the biggest of thanks are in order for my non-academic friends and family, who have kept me sane and supported me before, during, and after writing this book. Mum, Dad, Samu, Deh, Dany, Mattia, Nicole, Paolo, Ste: thank you for making all of this possible. I love you all so much. Zio Egidio and zia Nico: you have and always will be my second parents, and I can't thank you enough for all your love and help. Giovanna: thanks for being a mentor, colleague, friend, and an example to look up to. Ale and Biondo: thanks for growing up together, being there *literally* from day one (or more precisely six and sixteen, respectively), and teaching me the meaning of friendship. Alice, Alessia, Alessio, Anto, Asto, Bene, Bione, Chiara, Dany, Davide, Diddy, Dipa, Elisa, Gianpy, Gio, Giulia, Giulio, Padu, Pippo, Simo, Stefania, Tia: you make my hometown worth coming back to. Ale, Angelica, Pablo, Gabri: you are the living proof that friendships can endure no matter the distance. Andrea, Edo, Danny, Dedde, Flavio, Giulia, Ilaria, Matt, Nella, Tanya: you make London my other hometown. The Rugged Bunch, especially Silvia and Fede: thank you for being the best coffee shop in London and for the magical oat flat whites that make writing happen! Vivy: thanks for being my love, partner, friend, and daily source of inspiration, drama, and laughter—and yes, thank you also for learning how academic referencing works just to check this book's bibliography (not once, but twice!!).

# Contents

# Introduction

On the evening of Sunday, 23 July 2006 the Israel Defense Forces (IDF) launched operation 'Web of Steel'. Part of the 2006 Lebanon War, this infamous operation would soon be deemed one of the 'major blunders in the war' (Harel and Issacharoff, 2008: 131) by the same Israeli authorities who sanctioned its deployment. Web of Steel concentrated on the seizure of the city of Bint Jbeil on account of its symbolic significance (Kober, 2008: 27–29; see also Galai, 2019: 299–303), and was explicitly devised to initiate 'a ripple effect on the rest of the Hezbollah strongholds in southern Lebanon' (Katz, 2006). Bint Jbeil had played host to a notorious victory speech given by the then Hezbollah chief Hassan Nasrallah in May 2000 following the Israeli withdrawal from its security strip in Lebanon. It therefore represented an ideal strategic objective that could enable the IDF to energetically stifle 'Hezbollah's emotions ... speed up the enemy's defeat and end the war' (Harel and Issacharoff, 2008: 138). On closer examination, the planning and design of the military operation, consistent with its grandiose and provocative title, show how a rhetoric of victory became the linchpin of Israel's military strategy. The IDF based its claim to victory on largely symbolic, rather than tactical, considerations.

In his 2000 speech, Nasrallah (see *Nasrallah victory speech*, 2019) proclaimed that he was 'offer[ing] this victory to our oppressed people in occupied Palestine and to the people of our Arab and Islamic nation ... from the liberated Bint al-Jbeil', calling the 'people in Palestine' to engage in a new Intifada to get their 'land and legitimate rights back ... just like in Lebanon'. 'Indeed this Israel', Nasrallah declared, 'which owns nuclear weapons and the strongest air force in the region, by God, it is weaker than a spider web'. The July 2006 operation, then, would demonstrate that Israel was, indeed, a web: not of silk, but 'of steel'. The IDF 'wanted to answer Nasrallah's iconic speech': Bint Jbeil would be brought back under Israeli control, and victory declared 'over Hezbollah as well as the Palestinians in one go' (Galai, 2019: 301). It soon became explicit, in other words, that the IDF designed the military operation—name included—with Nasrallah's victory speech as their explicit point of reference.

*Military Victory Beyond the Battlefield*. Mirko Palestrino, Oxford University Press. © Mirko Palestrino (2025).
DOI: 10.1093/9780198955818.003.0001

Five days ahead of the beginning of the operation, Major General Binyamin Gantz reported to the IDF Chief of Staff Dan Halutz his intention to dispatch to Bint Jbeil, alongside fighting soldiers, a film unit that could 'record the story to the end' (Gantz, in Harel and Issacharoff, 2008: 138). The original orders that troops received, however, differed: the operation was to be 'limited to 48 hours' and they were not expected to capture the town: '[y]ou are going in, killing as many terrorists as you can and then coming out' (Adam, in Harel and Issacharoff, 2008: 139; see also 138). But Web of Steel ended up lasting much longer. According to Harel and Issacharoff (2008: 137–180) a host of disparate factors were responsible for the protraction of the operation, including the failure of peace talks following the Qana massacre in late July, Hezbollah's counteroffensive in Bint Jbeil, and unclear, contradictory command directives.

Undoubtedly, however, the symbolic considerations at the heart of Web of Steel regained the upper hand towards the end of the conflict (Harel and Issacharoff, 2008: 175–176). On 6 August, among a new wave of vague orders, a delayed operational schedule, and sporadic 'tense conversation[s]' (Harel and Issacharoff, 2008: 140) at the highest ranks of the IDF, Israeli infantry units had already passed the town when, equipped with cameras and other recording devices, they were ordered back into Bint Jbeil to 'film the planting of a flag on the site of Nasrallah's speech' (Galai, 2019: 301). Web of Steel would finally demonstrate the robustness and strength of the infamous 'spider's web' by capturing footage of the Israeli flag being raised over Hezbollah's stronghold, on the very same balcony from which Nasrallah delivered his speech six years earlier.

As Halutz would declare a few days after the launch of the operation, '[c]apturing Bint Jbeil has no tactical significance but it does have a symbolic meaning' (Halutz, in Kober, 2008: 28). According to the Winograd Commission,[1] the IDF command was patently open in identifying the production of a 'symbolic victory' (Winograd Commission, in Galai, 2019: 301), specifically in the form of a 'victory image' (Galai, 2019: 302), as one of the operation's paramount military objectives. For several commentators, the triumph the IDF was looking for in Bint Jbeil was not a tangible military success, but rather a 'staged' one. In the words of the head of the IDF's Strategic Planning Division, for example, the Israeli military only aimed at a 'victory staging ... [a] simulat[ion of] a state of victory' (Gilady, in

[1] The Winograd Commission is a specialist group appointed ad hoc by the Israeli government to enquire into the 2006 Lebanon War. The commission was composed of a retired judge, two retired IDF major generals, and two academics who were asked to draw lessons learned from the conflict (see Galai, 2019 for more details).

Galai, 2019: 300). Amidst a war that was, and still is, 'commonly viewed as a failure' (Galai, 2019: 299), a victory, even if staged, was believed to be enough to produce a 'real' one. In fact, the production of a victory image would unmake the ongoing military defeat and prove decisive in bringing the hostilities to successful cessation (see Katz, 2006; Harel and Issacharoff, 2008: 176; Galai, 2019). As IDF Brigadier General Eival Gilady (cited in Kober, 2008: 34) put it: '[i]t became a matter of producing a victory show'.

Ultimately, Web of Steel failed to deliver the glorious 'victory image' it was expected to produce (Galai, 2019: 303). The operation would go on to be received by the Israeli public as a military failure (Harel and Issacharoff, 2008: 131). Yet, the alluring promise of a 'victory image' that could quickly bring conflicts to a close remained so central to the IDF vocabulary and strategic thinking, that operations Cast Lead 2009, Pillar of Defence 2012, and Protective Edge 2014, and the 2018 and 2021 operations in Gaza, retained the informal objective of capturing an equally valid victory trope (Galai, 2019: 305–307; 2021). But how can a military failure be turned into a victory? And why is an image considered to be central to this shift? How can an image, amidst aversive military developments, bring an end to a conflict?

## A victory problem

Conventional wisdom has it that military victory usually comes at the end of war, as a result of the demonstrated combat superiority of one warring faction over the other. At Bint Jbeil, however, the IDF seemed to believe that video footage, or an image, could work just as well. Emerging from operation Web of Steel, in other words, is a vivid and complex puzzle that both resembles and exceeds a long-standing 'victory problem': what brings conflicts to closure? What determines victory in war? And ultimately, what does *victory* mean?

The most popular answer to these questions is that victory is both an opaque and undertheorized concept: competing notions of military victory can be identified across different cultures and in different intellectual traditions, throughout history (see, for example, Martel, 2011a, 2011b). The scholarly solution to this problem is to differentiate between levels of war and analysis: superior tactics and successful military operations lie at the basis of military victory, whereas skilful politics are the key ingredient for strategic victory (Mandel, 2006, 2007; see Bartholomees, 2008; Zaidi, 2009; Martel,

2011a, 2011b).[2] 'A classic error in warfare is "to mistake military victory for political victory"', Robert Mandel (2007: 467) writes, 'and ignore the reality that "victory is not assured when the shooting stops". To be counted victorious, a leader has to progress beyond military triumph to preserve the political control needed to secure an advantageous and enduring peace settlement'. Strategy, tactics, and other 'military sciences', it follows, should focus on producing opportunities for politicians and statespeople to exploit and translate into long-lasting, 'strategic victories' (see Mandel, 2006, 2007; Bartholomees, 2008; Martel, 2011a, 2011b).

Web of Steel, however, transcended tactics, strategy, and politics. In Bint Jbeil in 2006, the IDF thought that a victory image could be that decisive factor that would make the fighting stop in the first place. Military action was undertaken for the sake of recording a rhetorically powerful piece of footage that could convey a victorious image of Israel. The emotional resonance that this footage was deemed capable of evoking effectively brought victory-related rhetorical considerations to the fore, while simultaneously subsuming and overshadowing tactical ones. Indeed, tactics were curbed in favour of a perceived pressing need to produce a symbolic success, and troops accordingly ordered back into the city in spite of a previous announcement that Bint Jbeil had already been secured (see Harel and Issacharoff, 2008: 137–143). Politics, in other words, informed strategy, which informed tactics. Casualties were suffered precisely on account of a strong conviction that an image sufficed to end the war.

On one hand, the battle of Bint Jbeil suggests that the 'victory problem' is still very much a vivid one: victory continues to be an unclear notion that, nevertheless, remains central to world politics. On the other hand, it also shows a certain inadequacy of popular distinctions between military and strategic victory, tactics and politics, and war- and peace-winning. While the IDF command probably shared the idea that tactical successes are not necessarily decisive, nor final, they also seemed to believe a picture could be. The right image, they thought, could not only end the war, but even rewrite the extant military record of the campaign. It was their solution to the victory problem.

[2] This distinction between military and strategic victory is only one of several existing typologies. There is an entire line of research that seeks to enhance victory's conceptual clarity by tailoring analyses and definitions to different 'levels of conflict'. Other typologies differentiate between tactical, operational, strategic, and grand strategic victories (Biddle, 2004; Zaidi, 2009); tactical, strategic, and grand strategic victories (Martel, 2011a, 2011b); tactical, operational, and strategic victories (Bartholomees, 2008); and operational, strategic, and political victories (Gray, 2002). The underlying logic discussed here is analogous across typologies. I discuss its temporal implications in Chapter 2.

This book rethinks what victory is, means, and does. It reframes the victory problem not as one of conceptual confusion or specificity, but as a *temporal* issue. When it comes to bringing wartime to conclusion, fighting and other military developments are a rather indeterminate temporal marker. Instead, the social construction of wartime, peacetime, and victory as the watershed moment between these two temporal phases is realized through a plethora of social processes that have little to do with warfighting, and everything to do with associating temporal closure to a specific military outcome—be it a victory, in the sense of demonstrated battlefield superiority, a stalemate, or even a defeat. The highly coveted victory images that the IDF seems to be strenuously trying to capture war after war are one of such social processes—what I refer to, throughout this book, as 'victory practices'. While responsible for the production of military victory as the outcome of armed conflicts, victory practices are not bound to the wartime battlefield and straddle the lines between war and peace, military and civilian, kinetic and non-kinetic.

The central argument running through the book is that victory is not one of the material outcomes of armed conflicts but a temporal, sense-making device. Victory is produced just as much outside the battlefield as on it, during both 'wartime' and 'peacetime'. In fact, the political and emotional pull of military victory rests precisely on its capacity to mould time while being perceived as a seemingly timeless, or atemporal, occurrence: a material fact, an event precipitated by military developments 'on the ground'. Notably, I do not simply claim that victory is a social construct. Rather, I argue that it is precisely *in* and *through* time—understood here as an ordering tool and not a natural element in the background of social life—that victories are made and remade, and that the (conceptual) politics of victory can be grasped.

To defend this argument, I analyse six different victory practices, namely doctrine-making, history-teaching, victory celebrations, videogaming, military tattooing, and the erection of victory monuments. Not only do I demonstrate that bringing wartime to an end is part and parcel of all these different social processes, but I also detail the formal temporal mechanisms through which victory is made into the end of war. I find that the relationship between victory and time is intimate, complex, and not immune to inconsistencies, fissures, and discrepancies. I therefore challenge hegemonic understandings of victory as a material, military event that comes at the end of conflicts. Victory can end war but can also extend it. It can unite troops or even entire nations, but also powerfully pit them against one another. To reduce victory to the outcome of conflict, then, is not only analytically questionable but also politically problematic.

## Victory still matters

An easy way to avoid these analytical and political difficulties would be to abandon the concept of victory altogether. Indeed, several scholars have repeatedly espoused this position, dismissing victory as analytically insignificant (Angstrom, 2007: 2–3; O'Driscoll, 2019: 901).[3] Following the transformation of war (see van Creveld, 1991) or various revolutions in military affairs, the standard argument goes, wars have become unwinnable and victory an ill-suited, anachronistic concept that has no practical meaning or analytical purchase (Martel, 2011a: 132–135; 2011b: 514–518; O'Driscoll, 2019: 901; 2020a: 110–112). Moreover, as Cian O'Driscoll (2020a: 7; 92) has recently pointed out, inherent in the notion of victory are 'connotations of decisiveness and totality' as well as a 'negative conceptual baggage'—such as the idea of conquest—that have led normative enquiry on war, among other strands of scholarship, to dispense with victory as a concept because it is antithetical to their research ethos.[4]

The main problem with this position is that it is not consistent with the empirical record. For starters, victory is ubiquitous in political discourse (Hom, O'Driscoll, and Mills, 2017: 2–4; Hall, 2022; Hartnett et al., 2022; see also Bond, 1996: 1; Blum, 2013; O'Driscoll, 2019: 901; 2020a: 4). Even the existence of a solid consensus that victory is conceptually fuzzy (e.g., Mandel, 2006: 13; Bartholomees, 2008: 25; Martel, 2011a: ix; O'Driscoll, 2020a: 4) has not discouraged politicians, military strategists, policymakers, and other practitioners from employing the term. On the contrary, 'victory abounds' and its use seems to be widespread across history, 'from antiquity to the age of Trump' (Hom, O'Driscoll, and Mills, 2017: 2; see also Hall, 2022). Indeed, victory's unclear meaning might even have fostered its use in political discourse. William C. Martel (2011b: 532–533), for example, noticed that politicians tend to use the conceptual ambiguity of victory to their advantage, skilfully employing the term as self-explanatory and thereby altering its meaning from time to time to best fit their political goals (see also Mandel, 2006: 3).

[3] As examples of this position see Hobbs (1979), Johnson and Tierney (2006), and Mandel (2007: 465–466). For critiques, see O'Driscoll (2020a: 110–126) and Hartnett et al. (2022).

[4] In other words, while the relevance and the importance of studying war as a fundamental social and political phenomenon often goes undisputed (see Barkawi and Brighton, 2011: 126–127), the same does not seem to apply to victory, whose status as a distinctive analytic category for the study of war—as well as a self-standing object worthy of enquiry—is sometimes compromised by the controversial language it evokes and implies.

Moreover, perceptions of victory still affects policymaking and military strategy (Martel, 2011a: 4–6; see also De Franco, Engberg-Pedersen, and Mennecke, 2019a: 891). This was especially clear in the battle of Bint Jbeil. The Winograd Commission's report on the 2006 Lebanon War concluded that military operations were sometimes guided by the 'conscious or symbolic meaning of success', citing Web of Steel as an example of missions specifically designed to 'create images' of victory (Winograd Commission, cited in Galai, 2019). This insight, however, is not specific to that operation, or even just the IDF's modus operandi. A study by Johnson and Tierney (2006: 12–20, 2007: 47–49), for instance, demonstrated that perceptions of victory tend to impact policymaking in at least three different ways. Firstly, victory affects political and military leaders' careers. Depictions of specific individuals as victors are capable of rallying abundant support around them. Conversely, association with a military failure might bring an end to a politician's career. Secondly, perceptions of victory directly shape both foreign and domestic policy *during* the conflict. Support for the armed struggle fluctuates depending on the perceived likelihood of desirable outcomes; this, in turn, might result in self-fulfilling prophecies stemming from a general sense of defeat, or injection of additional resources into the war effort when a victory is felt to be on the horizon. Lastly, perceptions of victory profoundly affect historical learning, with models perceived as successful being replicated in the future, and failures being dismissed. Victory is also frequently mobilized in the biographical narratives of collective subjects, especially states (Subotic and Steele, 2018; Galai, 2019, 2021; Hall, 2022),[5] and remains central to our understanding of war as a phenomenon (O'Driscoll, 2020a: 6; see also De Franco, Engberg-Pedersen, and Mennecke, 2019a: 891–895).[6]

Unsurprisingly, then, not everyone endorses abandoning victory. '[G]allons of ink have been expended over the centuries on how to win wars', writes retired US army infantry colonel and Military History and Strategy professor J. B. Bartholomees (2008: 25), adding that nevertheless, defining victory remains 'the biggest theoretical challenge facing security

[5] Often because it is an 'emotionally charged' (Mandel, 2006: 13) and 'rhetorically powerful' ((Mandel, 2006: 13) and 'rhetorically powerful' (O'Driscoll, 2020a: 4) notion (see also Engelhardt, 2007; Scheipers, 2017: 48) that holds 'the potential to stir the *libido dominandi* of political and military leaders and to induce hubris' (O'Driscoll, 2020a: 128–129). I discuss this affective dimension of victory in more detail in Chapter 2 and Chapter 5.

[6] This was especially evident in Bint Jbeil. According to Galai (2019: 300), the 2006 Lebanon war 'was widely supported in Israel, much because of the unflinching endorsement from the media'. In turn, the idea of victory in general, and of the 'victory image' in particular, was central to the Israeli mediascape: via victory war was made tangible, achievable, and arguably justifiable (see Galai, 2019: 300–303).

professionals today'.[7] Accordingly, many social scientists[8] have taken a different route: scholars have explored the victory problem far and wide, generating, for example, context-specific genealogies (O'Connor, 1969; Hobbs, 1979; McCormick, 1986; Bond, 1996; Howard, 1999; Whitman, 2012; Strachan, 2016), overarching typologies (Gray, 2002; Mandel, 2006, 2007; Zaidi, 2009; Krause, 2019), as well as theories of victory (Biddle, 2004; Johnson and Tierney, 2006; Bartholomees, 2008; Martel, 2011a, 2011b) that seek to dispel the conceptual ambiguities affecting the term.[9] Substantial work has also been conducted on victory's normative and ethical dimensions,[10] and on 'determinants' of victory, especially in asymmetric warfare and civil war settings.[11]

[7] Several scholars agree on this point that victory research is underdeveloped (see Mandel, 2006: 2; Angstrom, 2007: 1–2; Martel, 2011a: 7; 2011b: 516; Blum, 2013: 392; O'Driscoll, 2015: 800–803). For example, while acknowledging that the incremental involvement of Western democracies in 'unconventional' warfare recently engendered a sharp increase in the scholarly production on the meaning of military victory (Martel, 2011b: 516), William C. Martel (2011a: 7) concurrently noted that a 'literature of victory that is distinct from theories of war, strategy and politics' is nevertheless lacking. Similarly, Mandel (2006: 2) highlighted that 'most discussions of victory are case-specific' and 'still decidedly underdeveloped'.

[8] This section focuses primarily on work conducted within the discipline of IR, broadly understood. It should be noted, however, that scholarship on victory does not necessarily conform to clear disciplinary boundaries, and that some of the studies considered throughout this book span academic disciplines and subfields such as Political Science, War Studies, Strategic Studies, History, Ethics, and Philosophy.

[9] While contributions to this research avenue are diverse, spanning extensive studies of single-level victories (Gray, 2002; Biddle, 2004) and broader multilevel analysis (Mandel, 2006, 2007; Bartholomees, 2008; Martel, 2011a, 2011b), they broadly agree on certain themes, namely the non-linearity of inter-level relationships (i.e., that tactical or even operational winning is not a necessary condition for strategic triumphs) and that victory is non-permanent, subjective, and political at all levels of conflict (Bartholomees, 2008: 26, 35; Mandel, 2006: 9–12, 2007: 466; Martel, 2011b: 531; Whitman, 2012: 251; Blum, 2013: 392). They also underscore the importance of distinguishing between conflict termination and conflict resolution (Strachan, 2016: 15). For the sake of this discussion, these contributions can be understood as a relatively coherent body of work concerned with clarifying what victory means and how it plays out empirically.

[10] It is important to note that most of these contributions locate and frame the victory question within the Just War Tradition (Wilcox, in Hartnett et al., 2022). While not immediately preoccupied with clarifying what victory means, this scholarship still offers valuable insights to grasp some of the sociopolitical implications that victory engenders—for example, distinguishing between military and moral victory (Scheipers, 2017: 48; see also Heuser, 2017; Patterson, 2017: 107–111) and *jus post bellum* and *jus post vim* (Brunstetter, 2017). For examples of this scholarship see Bellamy (2008), Blum (2013), Heuser (2013), O'Driscoll (2015, 2019, 2020a, 2020b), Rossi (2015), Brown (2017), Eckert (2017), Hom, O'Driscoll, and Mills (2017), Johnson (2017), Tierney (2017), and Hartnett et al. (2022).

[11] While some studies in these literatures remain concerned with questions of meanings and definitions in a way that closely resembles the work on interstate warfare mentioned above (see, for instance, Angstrom and Duyvesteyn, 2007; Brunstetter, 2017), the most conspicuous strand of this scholarship bypasses, when not explicitly dismissing, the definitional issue *tout court* to answer, instead, the question of *how* victory can be achieved by means of statistical analysis. For example, following David Lake's (1992) pioneering article, a substantial body of IR scholarship has contributed to what is nowadays labelled 'democratic efficacy literature' (Choi, 2004: 663) or 'democratic victory thesis' (Henderson and Bayer, 2013: 303)—a literature that maintains that the likelihood of winning wars is greater for democracies than non-democracies (Lake, 1992, 2003; Desch, 2002, 2003; Reiter and Stam, 2003; Choi, 2004; Downes, 2009; Gibler and Miller, 2013). For examples of scholarship on victory in asymmetric warfare see Rotte and Schmidt (2003), Angstrom and Duyvesteyn (2007), Sullivan (2007, 2008, 2012), Hasian (2010), Peic (2014), Sullivan and Karreth (2015), Tierney (2015), Brunstetter (2017), and Mills (2017). On civil war,

In short, victory is still central to both international theory and practice. If anything, then, victory research is doubly urgent. On the one hand, the centrality of victory to identity narratives amplifies the call to grapple with a notion that is not only prominent in public discourse, but that also seems to transcend supposedly different realms such as wartime and peacetime politics. On the other, it also reveals that to keep confining victory research to war politics—be they tactics, strategy, or conflicts' ethics—is to miss that victory plays a crucial role beyond them. Victory might be vague and hard to grasp, but it is hardly an anachronistic, irrelevant, or dismissible idea. In Bint Jbeil, victory enabled even more violence. Empirical relevance and theoretical elusiveness can be taken as calls for more, rather than less victory research, but also as an invitation to stop looking for victory only in the most obvious places.

## Victory and the end of war

Indeed, being bound to the wartime battlefield, hegemonic understandings of victory as the end of war serve to make war an *exceptional* social activity with clear beginning and endings. Victory has long been understood as a dot on a timeline, a watershed moment that marks the end of conflicts and separates wartime from peacetime (Hom, O'Driscoll, and Mills, 2017: 7; O'Driscoll, 2020a: 4; see also Gagnon, Senders, and Brown, 2014: 1–5). According to Cian O'Driscoll (2020a: 113), for instance, 'victory has historically been understood as closely associated with—or even predicated upon—the defeat of the enemy in so-called "decisive battles"'. While such an association is arguably the result of bad historiography—to the extent that it is much more 'mythical' than 'real'—its embeddedness in Western thought is so deeply rooted and firmly anchored that it 'still bears significant normative clout' (O'Driscoll, 2020a: 115).[12]

References to victory as a temporal marker or turning point are still ubiquitous in contemporary politics. Addressing the US Congress on 21 December

see Gent (2008), Cunningham, Skrede Gleditsch, and Salehyan (2009), Jones (2017), and Quinn, Joshi, and Melander (2019), Ryckman (2020).

[12] Interestingly, for James Whitman (2012) this link between victory and finality is not mythical per se, but should be traced back to legal, or ideological, imaginaries rather than military, or technological, ones. Rather than the increased deadliness of weaponry and other military technologies, for Whitman (ibid.) the recent decline in clear-cut victories is due to the changed legal or moral landscape, and in particular the weakening of *just victoriae*. In the past, he explains, pitched battles were widely recognized as a legitimate tool to settle disputes—something akin to a form of trial or legal proceeding—and victory worked as the ultimate judge, marking the end of battles, wars, and disputes. Whilst clearly a sociolegal argument, Whitman's approach points in the same direction as mine: victory practices that take place outside the wartime battlefield still shape the link between victory and time.

2022, Ukrainian president Volodymyr Zelenskyy wished for '[v]ictory. Only victory', and therefore asked for more weapons and support to finally lead 'the Russian Army to pull out completely' (Goncharova, 2022). During the 2015–2016 presidential campaign, Donald Trump's rhetoric hinged precisely on the 'reassertion of winnability': continuous reinstatements that the victories he would deliver would be 'proper' ones, of the kind that would finally put an end to 'endless wars' (Hall, 2022: 7–9; see also 2024). Notably, invocations of this kind are by no means exclusive to Trump's right-wing political discourse. Fanon's decolonial project, for example, is predicated upon a similar understanding of victory. In *The Wretched of the Earth*, indeed, the idea of victory signifies both a decisive event that 'break[s] the back of colonialism' (Fanon and Philcox, 2004: 146), and a time-altering turning point at which the colonized is reborn to a 'new man [who] knows that his life as a man begins with death' (Sartre, in Fanon and Philcox, 2004: lvi–lvii)[13]—so much so that, despite knowing that 'others will profit from the victory [and] not him' (ibid.), the colonized still 'sharpen their weapons to secure [a] victory' (Fanon and Philcox, 2004: 8) that will usher in a new era.

This univocal understanding of the link between victory and time, however, does not stand up to empirical scrutiny. Rhetorical invocations that the line between war and peace has become blurry, or that victories are increasingly undecisive, are but a recent example of this problem. Different national historiographies, for example, situate victory and the end of the First World War at different moments, years, and even decades (Gerwarth, 2016; see also Gerwarth and Horne, 2012; Palestrino, 2025). According to Tarak Barkawi (2016: 201), war has been a constant presence for the past five centuries in most of the world, and 'dominant historical periodisation of major wars as interruptions of the peace' only make sense in, and for, the Global North. From a postcolonial perspective, major victories such as 1918 and 1945 did not end war, nor bring about peace (see Chapter 1).

If victory is supposed to *end* war, how come even when military superiority translates into victory, war does not necessarily end? And how can some conflicts be over even in the absence of a clear military victory? Here lies the temporal version of the victory problem: much of the confusion surrounding the notion of victory concerns its relationship with time. Indeed, while plenty has been written about victory, scholarly production on the

[13] This reading of the particular temporal orientation of Fanon's decolonial project finds resonance in Homi Bhabha's (in Fanon and Philcox, 2004: xvii) suggestion that decolonization, for Fanon, is predicated upon a 'double temporality' that implies both a day-to-day commitment to the liberation struggle and 'a project of futurity' (ibid.) that depends on escaping the Manichean logic of the Cold War.

link between victory and time is in short supply (for an exception see Hom and Campbell, 2022b). This is not to say that time is nowhere to be seen in the prolific literature on military victory. Quite the opposite. According to Robert Mandel (2006: 13), for instance, to better grasp the meaning of victory we should split it, conceptually, 'into two highly interconnected yet distinct time phases' (Mandel, 2006: 13), namely 'war-winning' and 'peace-winning'. For him, that is, the difference between what others have called 'military' and 'strategic' victory is ultimately a temporal one: the first kind of success concerns combat and leads to the end of the hostilities, the second involves politics and leads to 'reap[ing] the payoffs of war', possibly 'yield[ing] durable postwar stability' (ibid.). On a similar note, Campbell and Steele (2017), suggest that the very idea of 'reasonable expectation of success' at the heart of Just War Theory is predicated upon a mythical rendering, or even idealization, of past victories as clear-cut events. An understanding of victory as temporal watershed, that is, is also central to Just War theorizing.[14]

Rather than the absence of time per se, the problem with extant accounts of military victory is a widespread tendency to treat time simplistically and leave temporal assumptions unchecked. Consider, for instance, the recent contributions to the Just War Tradition concerned with the body of moral precepts known as *jus post bellum*. This work is premised specifically on the idea that victory separates wartime from peacetime (see, for instance, Iasiello, 2004; Bellamy, 2008). Extending the traditional set of tenets of *jus ad bellum* and *jus in bello*, scholars in this subfield analyse belligerents' moral responsibilities *after* the cessation of hostilities—that is, once victory has been achieved. Notably, it is specifically the temporal switch engendered by the termination of organized violence—i.e., by victory—that brings about the change in moral regime and moral thinking (Iasiello, 2004: 37–40). Notwithstanding this centrality of victory to their moral reasoning, *jus post bellum* theorists devote their reflections to post-victory settings, avoiding engagements with the notion of victory altogether, as they employ it as a taken-for-granted starting point from which to develop their enquiry (O'Driscoll, 2020a: 9–10). Uninterested in how victory comes about or what it means, this work reinforces the idea that victory is an event that clearly separates temporal phases

[14] Analogous temporal assumptions are common also among strategic and historical scholarship on military victory—such as the ideas of 'quickness' as part of decisive victories in Napoleonic warfare (Bond, 1996: 28–43) or 'quickness of adaptability' in protracted wars of attrition such as the First and Second World Wars (Hobbs, 1979: 59–74, 173–176; see also Bond, 1996: 104–173), as well as the importance of 'timeliness' and 'drawdown timelines' for the exit strategy of contemporary counterinsurgency interventions (Kolenda, 2019; see also Mandel, 2006: 97–112).

and associated moral regimes.[15] The aforementioned distinctions between war- and peace-winning, or conflict termination and conflict resolution (Strachan, 2016: 15; see also Mandel, 2007: 466)[16] are another example of these dynamics.

In short, the uncritical acceptance of the idea that victory ends war does not only concern politicians and other leaders but also a large volume of academic output on military victory. My aim here, though, is not to dismiss this body of knowledge. Instead, I am concerned with the sociopolitical consequences of reproducing a hegemonic understanding of victory as the end of war.

## Victory, binaries, and chains of violence

Just like work on *jus post bellum*, hegemonic understandings of victory reproduce a stark differentiation between war and peace. As feminist scholars have explained time and again, however, this distinction is not only inaccurate but also deeply problematic. Violence does not abide by abstract binaries such as public/private, personal/political, or international/domestic. It unfolds, instead, in chains and continua (for an overview, see Wibben, 2020; and Yadav and Horn, 2021).[17] '[A] gender analysis', writes Cynthia Cockburn (2004: 43), 'suggests that it is meaningless to make a sharp distinction between peace and war, prewar, and postwar'. Gender-based violence endures steadily across all these phases, so that '[t]o consider one moment in this flux in the absence of the next is arbitrary' (ibid.; see also Moser,

[15] This problem is not exclusive to Just War Theory. Studies in transitional justice—especially a recent turn concerned with post-conflict transitions (see Obel Hansen, 2019)—are also premised on the temporal differentiation that victory enables, for example. This field has recently lamented the emergence of a certain ambiguity surrounding the applicability of specific accountability measures 'as the lines between "war" and "peace" have become increasingly blurred' (Obel Hansen, 2019: 961). Essentially, transitional justice processes are now facing the same issue of legal regimes applicability other jurists and philosophers of law outlined with reference to martial courts' jurisdiction and executives' emergency powers (Dudziak, 2012: 36–40). The problem in both cases revolves around the temporal distinction between wartime and peacetime that victory sanctions, for it is specifically such a technical distinction that makes it possible to identify the legal or moral regime that should be enforced.

[16] This idea has been expressed in various ways and terminology often reflects a differential emphasis on specific aspects of it. For example, conflict termination and resolution often point at the distinctive ethos of these activities or suggests that war studies and strategy are concerned with the former, while peace studies, diplomacy, and politics deal with the latter. As detailed in the following chapters, however, the key idea I am interested in here is a central assumption underpinning all these distinctions, namely that a clear, temporal division can be introduced between war and peace—the former encompassing systematic fighting or heightened hostilities, and the latter the absence of it.

[17] Carolyn Nordstrom's (2004) notion of 'the-time-of-not-war-not-peace', as well as Robert Nixon's (2011) idea of violence's 'temporal overspill' are similarly relevant here (see also Krystalli, 2021, 2024). On the idea of a continuum of violence see also Krause (2015), Wibben et al. (2019), Krook (2020) esp. chapter 11, and Wibben and Donahoe (2020).

2001).[18] By keeping up these binaries we fail to recognize how 'processes of violence replicate themselves both temporally and spatially' and that 'there is no "entry" or "exit" from violence, nor is there a linear progression of violence' (Yadav and Horn, 2021: 7).

The conceptual shift from binaries to continua, in other words, is also vital from a political standpoint—or to use Cockburn's (2004: 24) words once more, it 'is not an optional extra but a stark necessity ... if we are to develop effective strategies for mitigating the effects of conflict and for restoring and maintaining peace' (see also Moser, 2001: 30). Both policies and analyses that religiously stick to the wartime/peacetime differentiation tend to neglect or even efface forms of structural violence that are only visible when considered on the abovementioned, multidimensional continuum (see, for example, Reardon, 1985). The invitation, therefore, is to reflect on our potential complicity in the reproduction of these dynamics. Indeed, dichotomous ways of thinking are so widespread and normalized that it is quite easy to uphold and reproduce them, even involuntarily.

For Chris J. Cuomo (1996), for example, a war/peace binary structures the great majority of existing work on the ethics of war. She writes: 'most discussions of the political and ethical dimensions of war discuss war solely as an event—an occurrence, or collection of occurrences, having clear beginning and endings ... But many of the questions about war that are of interest to feminists ... cannot be adequately pursued by focusing on events' (Cuomo, 1996: 30). For her, there are two sets of problems with what she calls 'crisis-based ethics'—that is, moral frameworks that fail to acknowledge that violence unfolds on a continuum. The first is that 'they distract attention from the need for sustained resistance to the enmeshed, omnipresent systems of domination and oppression that so often function as givens in most people's lives' (Cuomo, 1996: 31). Second, they 'might actually keep resisters complacent about and complicitous in the general presence of global militarism' (Cuomo, 1996: 32). To acknowledge the existence of chains of violence, in other words, is but the starting point of a feminist politics, for there are forms of violence that are systematically present outside of the wartime battlefield, that we not only fail to see, but with which we also risk being complicit.

It is precisely this double-sided problematique of negligence and silent reification that lies at the heart of this book's response to the victory problem.

[18] Arguably popularized by Cockburn, the notion of a 'continuum of violence' was first introduced 'by Liz Kelly in 1988 to describe the normalisation of daily violence leading to sexual violence' (Yadav and Horn, 2021: 3). However, feminist arguments on the unsuitability of the war/peace distinction—among others—are much older and can be traced back to at least 1938 (ibid.). Of particular note, in this regard, is Betty Reardon's (1985) *Sexism and the War System*, arguably one of the first books to offer a sustained analysis between gendered power structures and social relations of war.

Existing studies of victory are premised on a narrow understanding of time as '*chronos*'—i.e., 'time as quantitatively measurable duration' (Hutchings, 2008: 5). They tend to assume, in other words, that time and victory are two separate things and that only the latter affects the former—a military victory stops the shooting, a strategic victory brings peace on (see Chapter 1). On one hand, by treating time as an 'ontological primitiv[e]' (Jackson and Nexon, in Hom, 2018a: 73)—a sort of physical, given entity with no social bearing—this body of work fails to acknowledge that time, including wartime, can be acted upon. As the example of Bint Jbeil painfully demonstrates, victory practices are constitutive of wartime and peacetime: the IDF actively tried to halt wartime with an image and *thereby* bring about victory. On the other hand, then, extant accounts of victory also buy into, and reproduce, a discourse of wartime as the exceptional interruption of peace. Through victory, that is, they also uphold the war/peace binary.

As I will discuss in greater detail throughout the book, the uncritical acceptance of victory's temporal qualities does not only lead to conceptual confusion. There are political stakes at play too. Conceiving of victory as a war-ending event serves, for example, to reproduce a Eurocentric category of analysis that is responsible for, among other things, the effacement of some forms of violence. Such political effects are far more urgent and pervasive than questions of analytical accuracy, or the status of the research record on military victory. Not only, then, ought victory research to ask how victory can mark the end of wars, or what makes victories final and time-altering, but it also must ask for whom and how. Most importantly, we need to remain attentive to our complicity in reproducing the temporal parameters normally associated with victory, treating scholarship as another victory practice.

## Victory practices: on rationale and method

This book reimagines victory, starting precisely from its link to time and temporality whilst avoiding the types of reification discussed earlier—such as the naturalization of time, or the war/peace binary. In the analysis that follows, I attend to a broader set of victory practices than those typical of the traditional research archive on the topic. Processes such as fighting, diplomacy, or peacekeeping can offer only a partial account of victory. In fact, the social construction of military victory as the end of war is mostly realized through peace declarations, victory marches, the erection of monumental complexes, victory utterances such as Trump's and Zelenskyy's above, or the much-coveted victory image that the IDF was looking for in Bint Jbeil.

By victory practices, then, I mean social processes through which the temporal parameters of military victory get produced, reinforced, or even challenged and altered. I focus on how these negotiations take place, adopting a sociologically thick approach that pays specific attention to who configures the relationship between victory and time, how, and why. Methodologically, this entails abandoning any commitments to researching victory on the wartime battlefield, looking instead at the social construction of victory when- and wherever this might take place.

Victory is a temporal sense-making device that operates within and without the wartime battlefield. From a methodological standpoint, my case selection strategy focused on contemporary victory practices and aimed at maximizing geographical breadth and empirical range. In particular, each chapter considers one, or a set of, very different practices that are both visibly distinct from one another as well as increasingly larger in terms of geographical scope—starting with the British military (Chapter 2), to then move to the broader British culture (Chapter 3), the Anglosphere (Chapter 4), and finally the global level (Chapter 5).

By contemporary, I mean practices that have taken place in the past thirty years or are still ongoing. Notably, these processes are constitutive of military victory in a set of chronologically distant conflicts, including the World Wars, the Cold War, and the War on Terror. For example, the textbooks, commemorations, and videogames discussed in Chapter 3 are indeed constitutive of victory in the First World War, but most of these are still used, celebrated, and played to this day. While perhaps puzzling, I take these chronological discrepancies—that is, dissonances between the publicly sanctioned starting and ending dates of a given conflict, and their continued (re)production outside the spatiotemporal coordinates that war supposedly took place in—to offer further evidence that victory should not be understood as one of the possible material outcomes of war, but rather a sense-making device and tool of temporal ordering.

The cases explored in the following pages, in other words, should be taken as empirical illustrations of my overarching theoretical claim about the relationship between victory and time. While my methodological eclecticism is purposeful, and the temporal dynamics discussed here might well be found to be at play across additional histories, geographies, conflicts, and victory practices, I do not claim that these are the only ways in which the temporal politics of victory play out. I also do not want to dismiss the valuable contributions of extant work in War Studies and adjacent disciplines discussed in the previous section. Rather, my aim is to complement these accounts by opening up the victory research archive both spatially and temporally.

Ultimately, this book demonstrates that the social construction of victory defies clean distinctions between military and civilian, or war and peace-winning because the temporal qualities of victory get produced and contested on an everyday basis, within and without war. It is therefore essential to engage victory practices meaningfully, and not as ancillary to, or by-products of, warfighting.

## Outline of chapters

Chapter 1 sets out the conceptual parameters of the book and offers a theoretical framework for the study of the link between victory and time. After grounding this enquiry within scholarly traditions concerned with the critical study of warfare, the chapter introduces 'timing theory' (Hom, 2018a, 2020) as a suitable analytical apparatus to come to grips with the research puzzle set out here. I theorize time not as a natural element, but the symbolic signifier of ordering relationships that humans establish to make sense of changing phenomena. As a temporal category, or 'timing indexical' (Hom, 2018a, 2020), victory is no different: it can be understood as a sense-making device that arranges and synthesizes phenomena, people, and events to make war intelligible. Building on queer-theoretical accounts of time, the chapter also develops timing theory to study both textual and extratextual temporal signifiers. This allows me to also tackle the seemingly material nature that hegemonic understandings of victory work with—i.e., existing imaginaries of victory as a *material*, military event that ends war. I show that by addressing the victory problem from a temporal standpoint—that is, attending to both narratives and symbols of victory, including human and non-human bodies, artworks and other cultural artefacts—key questions that extant accounts of victory struggle to answer can be tackled, and that metatheoretical analysis of these same studies can ensue.

The following four chapters apply this framework to offer detailed analyses of contemporary victory practices and associated timing operations, across contexts. I have arranged them in scalar order to mirror the expansive geographical scope outlined above. Accordingly, Chapter 2 starts with the narrowest geographical level considered in this book: a national military institution. The victory practice considered in this chapter is the most recently published military doctrine in the UK. I find that even fighting manuals acknowledge that military victory transcends the wartime battlefield. Doctrinal sources, indeed, propose a 'nested model of success' as the military answer to the victory question. Within this model, political narratives play a crucial role, to the point of even altering 'facts on the ground' and proving 'decisive'. The chapter contextually shows that in addition to explaining how to win wars, this model is also employed to tap into the affective dispositions of a wealth of audiences—including the UK military, the public, and its enemies. In short, via victory, specific temporal and affective politics are deployed which are key to the war effort and its sustainment, within and without the battlefield. Therefore, the chapter argues that the relationship between military developments and the narratives articulated to make sense of them—or victory's material and narrational dimensions—should be understood as co-constitutive. On this account, victory is not the outcome of conflicts, but of knowledge production—and timing—processes that are both narrational *and* material. Victory research, it follows, should not be constrained to the battlefield, but concomitantly explore co-constitutive victory practices that exceed it.

Chapter 3 moves the enquiry up beyond the military level, looking at the production of victory in UK culture. It does so through the analysis of a set of interrelated victory practices that are prominent in the UK: seemingly peaceful commemoration practices, history teaching materials, and videogaming. I first attend to GCSEs and AS/A Level history textbooks to show that victory is often taught as a 'thing of the military' also to civilians. In particular, textbooks attribute finality and causality to military developments alone. I then turn to Armistice Day rites to show how facts on the ground were far from being decisive in 1918, and that celebratory rites were crucial to engineer victory in the First World War. In fact, they still are. Despite a change in emphasis from victory to remembrance, the public rituals that were created in the late 1910s and early 1920s are still being observed today and testify to a victory that still hinges on its (over a century-long) continued reproduction. The chapter ends with a discussion of war videogames, another genre of cultural artefacts through which the (temporal) logics of victory are reproduced. Not unlike history textbooks or commemoration rites, war-themed videogames narrate and time victory as the direct consequence of facts on the ground. Overall, the chapter demonstrates that the everyday production of victory is not only geographically but also *temporally* dispersed—so much so, that Britons are still routinely engaged in winning the First World War.

Chapter 4 scales the inquiry up again, at the level of the Anglosphere, and attends to a recent, understudied, and yet popular victory practice: military tattooing. In particular, I analyse four recent exhibitions of military tattoos in the UK, US, and Australia. Homing in on individual war stories and their institutional re-mediations, the chapter disputes the idea that victory necessarily brings conflicts to an end. Treating tattoos and the bodies of their bearers as temporal clues, the chapter finds that temporal inconsistencies arise between individuals' war experiences and the war narratives that the museum articulates through them. Reinscribing their encounters with war within the contours of their autobiographical narratives, veterans employ tattoos and wounds as tangible traces of a war experience that is often unfinished or, as I suggest, *chronic*. As a result, these narratives are often silent about victory—the 'heroification' of these veterans in museums' openly celebrative accounts notwithstanding. Instead, mobilizing those same bodily traces in support of their war narratives, museums recount conflicts that have clear beginnings and endings and abide cleanly by the traditional wartime/peacetime distinction, offering *chronological* renderings of war experiences that silently efface *chronic* ones. Overall, the chapter shows that as soon as we stop looking for victory in the usual places, a plurality of wartimes comes to the surface. The hegemonic understanding of victory is exposed as just that: one among many possible articulations of the relationship between victory and time. To just assume that victory brings wars to an end is therefore problematic. It can mean being unfaithful to, and even effacing, some lived experiences of war. Instead, victory utterances that refer to temporal finality should be examined up close, paying attention to how victory is made to end conflicts and the political implications that ensue.

Finally, Chapter 5 turns to the global scale and attends to a last, widespread victory practice: monument building. I analyse three examples that are relatively distant from a spatiotemporal point of view: Baghdad's Hands of Victory, the Bronze Soldier of Tallinn, and Bolzano's Victory Gate. These monumental complexes are also architecturally different and attracted dissimilar levels of contestation. Nevertheless, they all buy into and amplify well-known tropes about the victory–time link in one way or another. The distinctive lesson learned here, is therefore threefold. First, victory is a popular and affectively resonant notion because it provides social orientation to both individuals and collectivities, especially when charged with temporal connotations. Second, the success of this emotional appeal is not a function of proximity or faithfulness to the historical military record: monuments can be successful even when they embody a fictional victory. Third, the social construction of military victory is not only realized through narratives and symbols that are *primarily* about victory: even episodes of contestation of a given victory symbol can reproduce the temporal dynamics that engendered that very contestation in the first place.

The Conclusion returns to the Battle of Bint Jbeil. After writing the epilogue of the IDF's quest for a victory image, I reread this empirical vignette against the backdrop of the arguments and findings presented in this book to tease out three main contributions that this work brings to IR, War Studies, and cognate disciplines—namely, a new ontology of victory, a method to study concepts *in* and *through* time, and a pair of theoretical innovations to timing theory. The chapter expands on each of these interventions, outlining both their contemporary relevance and some avenues for further research that stem from them.

# 1
# Timing Victory, Timing Warfare
## Narratives, Symbols, and the Critical Study of War

How is a military victory turned into the end of war? In the Introduction I showed that the abundant existing literature on victory does not offer direct answers to this question, nor the tools to come to grips with it. Accordingly, this chapter devises a conceptual framework that addresses the temporal version of the victory problem. It does so by turning to recent critical work on time and temporality in International Relations and adjacent disciplines. I examine the link between victory and time by treating references to victory as 'timing indexicals': clues that temporal operations of ordering are going on behind the scenes. This is the case for both oral and written narratives of victory, as well as textual and extratextual victory symbols. An exhortation to win war, a scholarly theory of military victory, or even an image and a triumphal arch can all signify victory and *time* through it. In other words, victory references conceal dynamic temporal processes that order people, phenomena, and events. I argue that victory research ought to attend to these ordering processes, and uncover who is carrying them out, how, and why. In this way, the temporal mechanisms underpinning victory become clear. So do their politics. This chapter sketches the contours of this conceptual framework for the temporal study of victory and applies it to a couple of recent, popular narratives and metanarratives of victory to illustrate how it works.

## Victory and (war)time

One of the ways in which military victory is turned into the end of war is precisely through victory utterances that subscribe to hegemonic understandings of this concept. Grasping this logic requires attending to the nexus between victory and time. Directly or indirectly, explicitly or implicitly, references to victory as the end of war reify, and therefore lend credence to, only one specific articulation of the relationship between victory and time:

*Military Victory Beyond the Battlefield*. Mirko Palestrino, Oxford University Press. © Mirko Palestrino (2025).
DOI: 10.1093/9780198955818.003.0002

the latter is a natural element and the former an event that just takes place within it. Neither of these assumptions is tenable.

First, there is nothing natural, or obvious, about the temporal contours of war.[1] As Christopher McIntosh (2016: 118; see also 2020a, 2022a) puts it, 'war is not necessarily something that "breaks out" or is "sparked" by events, but rather a series of events and uses of force that get interpreted ... into one broader, linked event. Dominant discourses and narratives shape how we read these multiple disparate events' (McIntosh, 2016: 126).[2] The very idea of wartime is one of such dominant discourses. According to Mary Dudziak (2012), for example, wartime is a skilful political construction that enables violence: a narrative frame, or an argument, employed by political elites to bend the rule of law and authorize exceptional violence. On this reading, wartime has very little to do with warfighting (see also Barkawi, 2016; Palestrino, 2025). To be sure, war often features clearly discernible periods of heightened hostilities, but fighting is not the main constitutive element of wartime—and most certainly not the only one. '[W]e imagine wars to be bound in time', writes Dudziak (2012: 15), 'but the American experience is to the contrary'. In the words of a very well-known slogan: 'America is not at war. The Marine Corps is at War; America is at the mall' (Murdoconline, cited in Barkawi, 2016: 202). Dudziak (2012) explains that the reason why life in the US can be peaceful while the country is formally engaged in a Global War on Terror that is often deemed 'permanent' (Darda, 2019; see also Filkins, 2008) is precisely because wartime is largely independent of fighting, and is determined instead by legal categories, political discourse, as well as movies and other popular cultural artefacts.

Second, and relatedly, victory is not *separate* from time but, instead, takes part in its construction. In a recent analysis of US wartime in the twenty-first century, Hom and Campbell (2022b) have suggested that victory can be thought of as precisely one of the ideas that helps normalize war as the

[1] While the relationship between victory and time is understudied, that between time and the broader phenomenon of war has recently received some level of scholarly attention. This especially the case in Critical War, Military, and Security Studies, three emerging fields of enquiry that often follow in the footsteps of the early feminist interventions discussed in the Introduction. Anti-colonial scholarship in these areas, for instance, has voiced complementary concerns around the war/peace distinction (Barkawi and Brighton, 2011; Barkawi, 2016: 205), while recent feminist research on 'war as experience' has started to attend to forms of violence beyond this binary (e.g., Sylvester, 2011, 2013a, 2013b; McSorley, 2013; Parashar, 2013; see also Brandimarte, 2023).

[2] Accordingly, McIntosh (2016, 2020a) insists that we examine war narratives and discourses from both practitioners and scholars, questioning the political value of the patterns of past, present, and future that they set up and promote. Engaging in a critical reading of Clausewitz, for instance, he shows that 'war is socio-politically articulated as exceptional, but ... relies upon universalizing a contingent set of ideas about war's temporal dimension', and that war, instead, 'takes place in, and produces, "its own" political time' (McIntosh, 2020a: 555).

exceptional interruption of peace. '[N]ational calendars, public commemorations, and team sports', they write, 'form a "liturgy of triumph" ... that help[s] institutionalize wartime victories in modern society' (Hom and Campbell, 2022b: 593). Indeed, celebrations of past victories and other related performances continue to circulate imaginaries of decisive victories—no matter how anachronistic these might be—and, as a result, they reinforce a popular understanding of wartime as finite, exceptional, and temporary (see also Hom and Campbell, 2022a: 525; McIntosh, 2022a).

This logic, however, can be extended further. On one hand, it is not only ritualistic acts that contribute to the reification and normalization of wartime via victory—all references to victory have the potential to affect wartime, in one way or another. Consider, for example, the public and specialized discourses around victory discussed in the Introduction. Open calls to make war winnable are not dissimilar from victory-related national festivities: they also reinforce the idea that wars are temporally bounded. And so do scholarly calls to distinguish between different types of victory occurring in distinct temporal phases, or between War and Peace Studies. On the other hand, the production of temporal imaginaries via victory is not limited to wartime—or, put differently, the link between victory and time also extends, for example, to peace and other times. After all, national biographies and other identity narratives often identify specific victories as crucial turning points, or even their origin.

To come to grips with the 'temporal work' that victory does, then, we not only need to look to both direct references to victory and more implicit ones, but also beyond wartime. In fact, the very first step is to abandon any ontological commitment to time as natural (see Introduction; see also Hutchings, 2008; Hom, 2018a, 2020). For example, in their account of how US wartime works, Hom and Campbell (2022b: 595) suggest 'to identify the practices of timing war—that is, the intersubjective ideas and processes that construct and produce war as a temporal expanse empowering certain agents and authorizing exceptional actions and political relations'. From a theoretical standpoint, that is, these authors work with a 'heterotemporal' (see Chakrabarty, 2008: 95, 239; Hutchings, 2008: 166–169, 172–176) understanding of time as sociopolitical construct. In particular, they apply 'timing theory' (Hom, 2018a, 2020), one of the most influential approaches to time and temporality in IR.[3]

[3] For (critical) literature reviews of the so called 'temporal turn' in IR (Berenskoetter, 2011: 664; Hom, 2018b: 305) see Chamon (2018) and Hom (2018b, 2020: 8–9). For examples of work in this tradition see Agathangelou and Killian (2016), Hom et al., (2016), Cohen (2018), Lazar (2019), Rao (2019, 2020), and McIntosh (2023) .

Timing theory is inspired by Norbert Elias's (1978, 2007) processual sociology and premised upon the linguistic shift from the substantive 'time' to its verbal form 'timing' that gives the theory its name. It hinges on the single, crucial tenet that (all) time(s) is(/are) socially constructed (see Hom, 2020: 60) and proffers a change in the epistemic posture towards time and temporality. On this account, indeed, time is not a natural element in the background of social life, but a pervasive political idea that is simultaneously produced and influenced by human activities. More precisely, timing theory postulates that time is an 'act of intellectual synthesis': 'the symbol of a relationship that a human group ... establishes between two or more continua of change, one of which is used by it as a frame of reference or standard of measurement for the other (or others)' (Elias, cited in Hom, 2018a: 72)—such as clocks' numerical intervals or changes in the position of celestial bodies. Far from being natural, time is a human act—and wartime an excellent case in point insofar as it 'knit(s) together a multitude of messy and malevolent practices into a cohesive and recognizable "time of war"' (Hom and Campbell, 2022b: 595). Indeed, the vernacular use of the substantive form 'time', Hom (2018a, 2020) explains, proves illusionary as it hides dynamic operations of timing, concealing, together with its verbal origin, the politics and social qualities time embeds.[4]

I argue that references to victory work in a similar way. Just as the noun 'time' hides the relational and dynamic nature of timing processes, victory appears to be a material event that just *ends* war, intuitively. But, for that to be the case, victory needs *to be turned into* the end of war. Victory is not simply an occurrence and, like time, also needs to be de-objectified and denaturalized (Hom, 2018a: 72). Drawing on timing theory is a particularly convenient way to do so. Indeed, timing theory does not only explain what time is and how it works, but also how to study it. In particular, Hom's (2020: 47) invitation is to 'treat all references to "time" and "temporality", along with temporal modifiers and predicates, as *timing indexicals*, or symbolic discursive markers of underlying timing efforts ... as snapshots of timing-in-progress'. For every temporal clue encountered in a text, for example, it is possible to identify who is doing the timing and why. Moreover, this analysis of social agents' 'will to time' (Hom, 2018a: 73) can be supplemented with a focus on the successfulness of their timing attempts (Hom, 2020: 37–43).[5] In

[4] As Elias notes, indeed, linguistic transmission 'tend[s] to place at the forefront of our attention substantives, which have the character of things in a state of rest' (Elias, 1978: 111). Variation and action, instead, are usually conveyed through attributes and verbs (ibid.).

[5] While Hom (2018a: 72–73) warns us about the self-reifying nature of particularly successful timing attempts—i.e., passive, as opposed to active, and supple, as opposed to brittle, timing processes (Hom,

the latter case, the focus would be on the nature and features of the continua of change being related, in particular the one identified as the frame of reference, or 'timing standard', for the others (Hom, 2018a: 72). Clock-time or the Gregorian calendar, for instance, are good examples of timing standards that have been so successful that they are commonly perceived as natural.[6] Importantly, Hom (2018a: 72) specifies that 'a timing standard need not be natural or consistent'. Instead, the frame of reference only needs to establish some sort of ordering rule, creating some broad coherence that makes the social relation (i.e., the timing attempt) work[7]—as is normally the case with victory and wartime.

When it comes to the study of victory, both these modes of analysis can be put to work. Indeed, this approach can help map out the ascendency of the current hegemonic understanding of victory according to which victory *is* an event that ends war, as well as the sociological underpinning of this formation. It is also possible to explain how victory works the way it does from a temporal standpoint—i.e., to show which social and temporal mechanisms it follows—and which political interests it serves.

## Timing theory, timing victory

From a temporal standpoint, victory practices can be understood as timing operations in which different events and phenomena are coordinated and synthesized. Victory has all the features of a timing indexical: a naturalized and seemingly self-explanatory reference to the end of war that instead hides intricate processes of integration and ordering between varied phenomena, events, and people. For example, when uttered as a desired or actual end state of a conflict, military victory is, at a minimum, being posited as the outcome of specific martial processes construed as war with clearly identifiable starting points. In turn, as McIntosh (2022a: 2)

2020: 37–43)—the identification and study of timing indexicals is always possible, no matter how effective the timing operations they signify and conceal.

[6] Other examples of timing standards include, but are not limited to, daily routines, (auto)biographies of both individual and collective selves, or even broad and general narratives like seasonal change. In other words, the array of phenomena involved (or better, timed) in timing operations is substantially vaster than the clock- or sun-related temporal analogies we have in mind when thinking about time and temporality. In fact, the very existence of clock-based and calendrical time as shared temporal imaginaries indicates nothing but the successfulness of those timing operations. Notably, timing theory also allows us to map out the different sociologies of these times and explain their successfulness.

[7] 'According to its overarching standard', Hom (2020: 35) explains, 'timing proceeds through further operations of integration and coordination' whereby the 'when-aspects' of events, people and phenomena are gathered and arranged.

suggests, the traditional understandings of war that these utterances conjure up are not neutral: they 'link together past, present, and future in a way that makes ongoing assassinations or decades-long military occupation appear not only justifiable, but unremarkable acts of a discrete time of war'. Victory references, in other words, entail and conceal the inclusion and exclusion of specific forms of violence—i.e., what and when counts as war and not (see Sjoberg, 2013; Barkawi, 2016)—as well as the synthesis and coordination of some events and phenomena to the detriment of others.

Victory may work as the timing standard for other continua of change, especially identity and war narratives (see, for instance, Hom, 2018a: 75–76; Hom and Campbell, 2022a, 2022b). For example, victory might be used to bring conflicts to closure or to mark the birth, rebirth, or demise of social subjects (see Chapter 3 and Chapter 5). Sometimes, instead, victory gets synthesized in other timing operations that adopt different standards—such as identity, institutional, or scholarly commitments—as their frame of reference. Here, the temporal parameters of the notion of victory are silently reproduced and go almost unnoticed, as the concept is deployed uncritically, for example, to explain other phenomena (see Chapter 2 and Chapter 4). Either way, approaching victory through a timing lens sheds light on the specific temporal politics of these processes and offers formal explanations of how victories get 'undone' over time, or how and why wars might end for some people, but not for others.

Indeed, this conceptual apparatus is not only comprehensive and economical, but it can also be scaled up. By treating victory as a timing indexical, it becomes clear that both narratives and metanarratives of victory are still timing attempts. After all, it does not matter whether a victory reference is found in a political speech, a historical treatise, or a theory of war: it still works as an indexical. This type of temporal analysis, however, exposes the differential sociopolitical effects of these temporal operations, shedding light on, among other things, the will to time (i.e., time-related agency)—or even just unwilling complicity—of different social actors. As a result, the approach proposed here simultaneously enables the type of reflexive engagement that feminist scholars working on the continuum of violence call for. Victory utterances *time* irrespective of the express purpose of their timing actor. Similarly, analytical categories that are inaccurate and exclusionary can be reproduced even involuntarily. Alison Howell's (2018) call for a concept of 'martial politics', for instance, offers important insights into the dynamics through which scholars are implicated in the reproduction of the war/peace binary—a notion that is central to the concept of victory too.[8] She argues that notions of militarism and militarization—two pivotal ideas at the heart of much feminist and other critically oriented scholarship on war and security—can reinforce the very analytical binary they seek to oppose. 'Embedded in "militarization"', Howell (2018: 118) explains, 'is a theorization of "before and after"—of movement from a non-militarized (or less-militarized) state to a militarized one'. As a result, and despite claims to the contrary, militarization fails to break with the war/peace binary,[9] working instead towards its silent reinstatement.[10] In short, some of the concepts that scholars employ to study war, such as militarization, end up framing specific understandings of the time of war. The same is true for victory.

## Victory narratives

While it has been argued that military victories are becoming less decisive (Hobbs, 1979; Bond, 1996; Whitman, 2012), victory utterances are all the more frequent in political discourse (Hom, O'Driscoll, and Mills, 2017; O'Driscoll, 2020a). A recent study by Jonny Hall (2022: 2), for instance, has shown that George W. Bush, Barack Obama, and Donald Trump have all repeatedly 'invoked the empathic language of "winning" and "victory"

[8] A comparable problem of subtle reifications animated recent discussions on the ontology of war (see, for instance, Hutchison and Calkivik, 2019). According to Astrid Nordin and Dan Öberg (2015: 393), for example, 'these debates have involved a perhaps surprising level of agreement … on an ontology of war as fighting' that obscures the lack of antagonism that characterizes contemporary conflicts. For them, today's wars are about targeting operations rather than battles, and should be grasped as 'an administrative model' (Nordin and Öberg, 2015: 406) that, by centring operational *pre-planning*, leads to 'a disappearance of encounters' (Nordin and Öberg, 2015: 405). They write: '[i]f we look closely at the temporal aspects of targeting we find that it … is strictly speaking taking place from the future … In this modelling the possibility of a human encounter in the here-now disappears, since "now" has been pre-planned days earlier' (Nordin and Öberg, 2015: 404–405). As a result, to (over)emphasize the importance of fighting and antagonistic contests is to partake in 'an extreme reification of war' (Nordin and Öberg, 2015: 407). By centring fighting, we risk superimposing and reproducing an idea about war that is seemingly detached from at least some ontic realities. Instead, these authors suggest 'to find new ontolog*ies* complementing the notion that war is fighting' and offer war-processing as one of such alternatives.

[9] By contrast, for Howell (2018: 118), martial politics shows that '"normal politics" is not overtaken by "militarization"; instead, martial relations inhere in liberal politics as they are enacted on those who are racialized, Indigenous, disabled, queer or otherwise constituted as a threat to civil order' (Howell, 2018: 118). While not formulated in explicitly temporal terms, a similar critique has been recently put forward by Millar (2022) under the register of 'discursive martiality'.

[10] This idea of martial politics is by no means uncontroversial and has attracted both sympathizers and critics. Taking up Howell on her invitation, several scholars have dug deeper into the *already existing* martial relations that underpin seemingly non-martial, or even peaceful, institutions (see, for instance, Axster et al., 2021; Gibson-Fall, 2021; Manchanda and Rossdale, 2021). Just like Howell (2018) shows with reference to the police and the university, these works seek to demonstrate the fundamental *continuities* between war and society, emphasizing, for instance, the war-imbricated origins of schooling (Gregory, in MacKenzie et al., 2019; Winter and Mills, 2020) and neuropsychiatry (Fletcher, Zubair, and Roche, 2022) among others. Pushing back on the idea of 'martial politics', other scholars have instead suggested that, while useful, the concept misses the martial intensities and ambiguities intrinsic to processes of *de*-militarization (Wegner, in MacKenzie et al., 2019) or that it assumes what war is *a priori* (Shah, in MacKenzie et al., 2019).

when discussing [the War on Terror]'—even though this rhetoric may be out of step with an underwhelming military record. According to victory scholar William C. Martel (2011b: 532), implicit or even tacit references to different levels of victory were intentionally deployed by the Bush and Obama administrations for the sake of mobilizing public support around diametrically opposed foreign policy strategies. Seeking to justify the invasions in Afghanistan and Iraq, George W. Bush 'framed victory in grand strategic terms' (ibid.), identifying 9/11 as the *casus belli* of a war that aims to eliminate terrorism. Conversely, Obama reframed 'victory in more strategic and one might even say tactical terms' (ibid.), arguing for demobilization in Afghanistan on account of the 'sufficient progress' made 'on the ground'.[11]

Far from merely shifting from one notion of victory to the other, I suggest that the differential uses of victory by the two administrations conceal much more complex timing attempts that, when examined up close, shed light on the subtle or hidden politics of both narratives. Indeed, conceived as a timing indexical, victory ceases to be the analytical descriptor of distinct processes, becoming instead the symbol of human-led, dynamic relations of coordination and synthesis between carefully selected events and phenomena. Who is doing the timing? What is their will to time? Do they want to bring the conflict to closure? How? Which event do they identify as the starting point of the war? What is the desired end state?

For example, in identifying 9/11 as *casus belli*, Bush's narrative frames the event as the beginning of a conflict that can have 'terror' as its designated enemy, and by extension, a longer time frame and a broader—in fact unlimited (see Chamayou, 2015: 52)—geographical scope. As a temporal clue, this victory is not only 'grand strategic' in its ethos to 'transform or realign international politics' (Martel, 2011b: 520); rather, it expresses a political will to wage a potentially endless and spatially unbounded conflict. In turn, to suggest that 'sufficient progress' has been made to warrant demobilization might well indicate the desire to bring a conflict to closure, but it also entails a complete reshuffling of the spatiotemporal coordinates of the processes under consideration, such that 'the conflict' under analysis is a rather different one. Here, Afghanistan can even be disassociated from the War on Terror—or at least conceived as a subset of it. As a result, a victory can be in sight if the starting point is not 9/11 and the desired end state is not the end of 'terrorism' but the 'pacification' of the country: the two narratives are

[11] For Martel (2011a, 2011b), shifts from one level of victory to another like this one are a convenient, and therefore recurrent, political strategy. For example, he detects similar shifts in the foreign policy strategies of different US administrations also for the Vietnam War (Martel 2011b: 532).

not necessarily politically incompatible, but they are undoubtedly temporally discrepant.

In addition to examining the sociopolitical positionality of the actor(s) engaged in these timing attempts, a timing lens also invites questions about the politics of choosing any given phenomenon as the frame of reference to time the others (Hom, 2018a: 75). Indeed, the use of grand-strategic and tactical notions of victory in the two narratives can be taken to work as differential timing standards: tactical successes marking the tempo of a narrative that seeks to urge demobilization, and grand-strategic victory working instead towards the spatiotemporal dilation of the War on Terror. Importantly, the politics of choosing one notion of victory over the other do not simply translate into more or less protracted conflicts. These narratives also identify, either explicitly or implicitly, what does and does not count as success: they *constitute* the fabric of that timing standard. In timing our daily routines through the clock, for instance, we endorse abstract numerical intervals as valuable standards against which to measure our everyday activities. Unlike clock-based or calendrical time, these victory narratives are a less clear and regular timing standard. Still, what is the equivalent of minutes and seconds when war is timed with victory as its standard? What, in other words, *constitutes* victory?

As he sought to make clear on different occasions, Bush's War on Terror would 'not end until every terrorist group of global reach has been found, stopped, and defeated' (Bush, cited in Hall, 2022: 5). Equating terrorism to the evils of Nazism and Communism and drawing repeated parallels between the War on Terror, Second World War, and the Cold War, Bush did not only try to articulate an ontologically secure (Hall, 2022) or temporally stable (Solomon, 2014) identity narrative that could reinforce the idea of American exceptionalism. Rather, Bush simultaneously elevated terrorism to an ideology, implicitly denouncing discrete military successes 'on the ground' as insufficient. Here, the discrepancy with the Obama administration's narrative is perhaps even starker. As Hall (2022: 6) notices, 'Obama restricted the securitization of terrorism ... to specific terrorist groups as part of the efforts to move away from the politics of Bush's "War on Terror"'. The result, however, is not simply a more 'winnable' war, or a more circumscribed spatiotemporal focus, but the upholding of 'tangible' military successes—i.e., satisfactory military operations—as the desired timing standard. On this account, that 'the Obama campaign was never particularly radical' and, as several commentators noted, resulted in 'a policy that emphasized killing' (McIntosh, 2022a: 12), is not surprising.

## Victory metanarratives

'The final observation', Martel (2011b: 533) suggests, 'is that when policy-makers use the word "victory" they refer to the level that best suits their political objectives and perhaps pay greater attention to their own domestic political audiences than to "objective" conditions in the theatre where the war was waged'. A temporal approach to victory, however, suggests they do much more than that. Consider the events and phenomena being related by the two different narratives from the Bush and the Obama administration above. Could we not say 'objectively' that even the grandest of victories in Afghanistan would be at best a necessary but still insufficient condition to end terrorism? Is it not 'objective' that—the aim being pacification—the absence of heightened hostilities constitutes some sort of 'progress'? The point here is not to highlight the pitfalls intrinsic to the very idea of objectivity, but the logical, and more specifically temporal, dependency that ensues from Martel's apparatus above—and from all typologies of victory, for that matter. If the problem lies, as Martel suggests, with the shift in the 'level of victory' under consideration, we could in fact consider both narratives to be objective *relative to their own standards*. Indeed, to denounce the absence of objectivity is to miss the point: each narrative is purposefully trying to uphold a different continuum of change as their preferred timing standard. And so is Martel's. To suggest a given strategy disregards the '"objective" conditions in the theatre where the war was waged' (Martel, 2011b: 533) is to contribute to the upholding of military developments, or 'facts on the ground' as the standard against which these narratives *should* be assessed—certainly a common practice, but not necessarily the only alternative (see Chapter 2).

By selecting a conflict's starting and ending points, as well as the specific events, processes, and phenomena to be related, coordinated, and synthesized, each of these temporal operations—Bush's, Obama's, *and* Martel's—establishes its own terms of reference: the timing standard that best suits one's political objectives. Martel's narrative, it follows, is not different: like Bush's and Obama's narratives, his theory of victory can be conceived as a timing attempt furthering temporal politics that we can grasp and examine. Through timing theory, in other words, the role played by metanarratives of victory can also be explored.

Martel's claim to objectivity, for instance, is intrinsic to the idea of scales or levels of war. However, if we treat victory as a timing indexical, the political implications of such seemingly innocuous analytical choices are brought to the fore. The risk with calling for increasingly specific descriptions of victory in the name of conceptual clarity, for instance, is to obscure the political agendas of timing attempts like Bush's and Obama's above. Moreover, the problem of reifying the war/peace binary, as discussed in the debates about militarization above, rears its head. As timing theory illustrates, by determining and sponsoring a specific timing standard, timing operations are fundamentally implicated in the design, reproduction, and normalization of (illusive, but very much real) temporal regimes—to the extent that today, seconds and minutes can be thought of as the epitome, if not synonyms, of time. The same applies to victory: to suggest that victory is the material outcome of martial engagements—be they individual or repeated—is to contribute to, or more precisely constitute, that outcome, or at least the illusion of it. A specific political agenda and intentionality—an explicit 'will to time'—are not necessarily at stake here: reification occurs whether we want it or not. In the same way in which we might not share the political commitments of Coordinated Universal Time (see Stevens, 2016), we still time our daily life through clocks and observe time zones, passively partaking in the ascendancy and reification of a 'Western Standard Time' (Hom, 2010, 2020: 9–11).

The vernacular, unreflective use of victory as the outcome of conflicts, for example, often attributes primacy to military developments over political narratives, while simultaneously obscuring the fundamental role that these narratives play in shaping those 'facts on the ground'—or materialities of warfare. If victory is conceived as the symbol of dynamic relations of timing in which a plethora of varied phenomena are related and synthesized, then both 'facts' and 'narratives' are constitutive of its meaning. Accordingly, as I will argue in Chapters 2 and 3, the relationship between victory's narrational and material dimensions should be understood as one of co-constitution rather than subordination—an analytical pitfall that is widespread among extant victory research and that has important political repercussions. Indeed, even accounts of military victory that emphasize that '[v]ictory in war is at the most basic level an assessment, not a fact or condition' (Bartholomees, 2008: 26; see also Howard, 1999; Johnson and Tierney, 2006) reproduce precisely the fundamental split between the ideational and material components of war that is responsible for generating the problem in the first place.

For example, Lauren Wilcox (in Hartnett et al., 2022: 3) recently argued that while the revival of victory research in Just War theorizing (see O'Driscoll, 2015, 2019, 2020a, 2020b; Hom, O'Driscoll, and Mills, 2017) is to be welcomed insofar as victory raises pressing ethical problems, 'this is not just a problem in "victory" but a problem in *war* and for thinking about violence and the structures of thought that surround it more broadly'. For her, to keep thinking about victory through the overly Western and abstract Just War canon, is to miss that war 'is resolutely bodily, violent, and destructive', that it 'exceeds the rational structures of control', and that, ultimately, '"[v]ictory" entails the transformation of bodily injuring and death into meaning and political power' (Wilcox, in Hartnett et al., 2022: 3–4). A temporal analysis of victory, I would add, sheds light on the mechanisms at the heart of this transformation. As proponents of martial empiricism suggest, however, we need to actively pay attention to 'the invasive ecology of warfare', in its capacity to target both 'the material *and* immaterial connective tissues of our life-worlds' (Bousquet, Grove, and Shah, 2020: 111).

## Victory matters: symbols, narratives and temporal traces

Normally, references to victory do not only imply the end of war, but also that victory is a material, military event. This material dimension of victory, I argue, is integral to its temporal politics. According to Bousquet, Grove, and Shah (2020), war is a distinctively material phenomenon, and yet, this materiality is rarely taken seriously. '[T]he privileged conceptual frames of discourse and identity', these scholars write, 'have accounted for the constructions of threats and figures of enmity, but they can say little about the inescapable materiality of war, de facto relegating it to a second-order effect of the former' (Bousquet, Grove, and Shah, 2020: 102). Their remedy, then, is to instead 'explore how war *becomes*' and focus on war's 'recurring persistence yet constant reinvention' (ibid.: 100). To do so, we ought to 'trace the lines of becoming that congeal into what comes to count as war' (ibid.: 101), and attend to 'the various bodies, objects, ideas, practices and affects that assemble to constitute collective violence' (ibid.: 103).

While treating victory as a timing indexical is one way to examine what counts as war, it is a strategy that does run the risk of disregarding the material dimension of war. Timing theory, after all, is a linguistic, discursive framework for the study of time, that is not immediately concerned with materiality. Methodologically, for example, Hom's applications of timing theory proceed mainly via textual, discursive analyses of timing practices (see Hom, 2016, 2017, 2018a, 2020; Hom and Solomon, 2016; Hom and Steele, 2016).

But time, too, has a material component. Paul Glennie and Nigel Thrift (2009), for example, open their volume on the history of timekeeping in England and Wales with an enticing introduction titled 'The Measured Heart' that vividly shows how clocks and bodies are more entangled than one might usually think. In their fascinating vignette, Glennie and Thrift (2009:

1–2) explain how Galileo, short of a mechanical timing standard, 'turn[ed] to human biology', and originally employed his own heartbeat to time the movements of the pendulum.[12]

As recent feminist (e.g., Loewen Walker, 2014) and queer theoretical work (e.g., Manchanda, 2015; Puar, 2017; Rao, 2020) on temporality show, one of the ways in which we grasp the political nature of time is by looking at the tangible traces it leaves on (human and nonhuman) bodies. According to Rahul Rao (2020: xx), for example, 'temporal movements', and in particular 'sensations of temporal lag and rupture', can have 'dramatic psychic and political consequences'. Engaging with both time and the body in even more explicit terms, Elizabeth Freeman (2010: xi) argues for a 'counterpolitics of encounter in which bodies ... meet one another by chance, forging—in the sense of both making and counterfeiting—history differently'. Freeman (2010: 3) recurs to the notion of *chrononormativity* to uncover the systems of mastery and (bodily) regulation enacted through temporal regimes:

> By 'time binds,' I mean something beyond the obvious point that people find themselves with less time than they need. Instead, I mean that naked flesh is bound into socially meaningful embodiments through temporal regulation: binding is what turns mere existence into a form of mastery in a process I'll refer to as *chrononormativity*, or the use of time to organize individual human bodies toward maximum productivity. And I mean that people are bound to one another, engrouped, made to feel coherently collective, through particular orchestrations of time ... Chrononormativity is a mode of implantation, a technique by which institutional forces come to seem like somatic facts. Schedules, calendars, time zones, and even wristwatches inculcate what the sociologist Evitar Zerubavel calls 'hidden rhythms,' forms of temporal experience that seem natural to those whom they privilege. Manipulations of time convert historically specific regimes of asymmetrical power into seemingly ordinary bodily tempos and routines, which in turn organize the value and meaning of time.

Not entirely dissimilar to timing theory, the idea here is that particularly successful timing operations come to be felt as second nature, 'like somatic facts' (ibid.). However, Freeman emphasizes the bodily dimension of timing, highlighting that what temporal regimes bind is the 'naked flesh' of 'individual

[12] References of this kind to time and the body, as well as to time and materiality, are sporadic at best within timing theory and, more broadly, IR work on time and temporality. In fact, social scientific work on time is often carried out via conceptual and discursive approaches that eschew material considerations to instead foreground a temporal politics that bypasses its material and bodily effects. For instance, Glennie and Thrift (2009) only employ Galileo's heartbeat example to introduce a discussion on time and modernity that does not confront bodies nor materiality.

human bodies'(ibid.).[13] The effects of temporal ordering, in other words, are not only registered in, and therefore visible through, language.[14] Time materializes both in language *and* on human and non-human bodies. It blemishes, scars, and even shapes bodies and material artefacts. For Freeman, those same bodies, subject to the binding of time, become tangible repositories of temporal traces and marks, and powerful, non-textual sources for the study of time, timing politics, and their sociopolitical consequences.

And if, as Freeman notices, via time bodies are made to 'feel' collectively, the agential power underpinning timing processes can still be uncovered. Indeed, the language of 'implantation' and 'inculcation' in the passage above exposes the politics of chrononormativity and raises questions about how 'institutional forces' can impose it (ibid.).[15] Freeman's approach, in other words, shows a clear, feminist commitment 'to paying attention to *systems* and *structures* of violence, as opposed to only examining individual physical manifestations of violence' (Krystalli, 2021: 16). It also embraces a feminist 'vision of embodiment in which bodies ... can be productive, as well as produced' and the 'materiality of bodies is not only an effect of political practices, but such practices are formed in relation to bodies as well' (Wilcox, 2015: 11)—i.e., it does not only uncover forms of institutional violence that defy the wartime/peacetime binary, but it also resists general tendencies to

[13] Here, chrononormativity finds close resonance with recent work in IR and germane disciplines concerned with rhythms and routines (Blue, 2019; Solomon, 2019; Wegner, 2021), as well as the social consequences of time and temporality—including accounts of 'time's structuring of particular social, material, and experiential inequalities' (Bastian et al., 2020: 292). Freeman's work also reflects critical accounts of collective identity formations (see, for instance, Solomon, 2014, 2015; Hom and Solomon, 2016; Palestrino, 2022a), especially in terms of collective coherence, or the idea of 'people ... bound to one another'. Her work, however, clearly indicates that time does not only bind selves, but also bodies.

[14] Importantly, I am not denying that language has a material dimension, that materiality is not linguistically mediated and can be apprehended outside of language or discourse, nor I am trying to introduce or reproduce a material/linguistic split. Rather, I am building on Freeman's work to suggest that if time does indeed operate extralinguistically—for instance, at the affective level—then an attention to extratextual sources is not only justified by a victory focus, or a critical war study ethos, but also by a timing approach itself. This certainly begs the questions of how to trace and approach explicitly extralinguistic effects of timing mechanism—one that I am much interested in, but that given this book's focus on victory, is beyond the scope of the present enquiry.

[15] Through the analysis of several literary texts, for example, Freeman shows how heteronormative familial time becomes normalized and (fictively and yet effectively) reinscribed within narratives of collective fate (Freeman, 2010: 21–33). As she explains, the transmission and normalization of the heteronormative familial model is realized through the circulation of photographic and cinematic media that embody familial memory tropes and aesthetics. The sociopolitical consequences of these processes, however, are not limited to the elevation of this model of kinship as the standard of a neoliberal 'progressive' teleology. Instead, individualized (and often gendered, raced, and classed) temporal ideals—such as men's or women's time—are simultaneously folded within. As a result, bodies are not only timed through the normalized idea of family. Contextual timing processes bind the 'men' and 'women' that are part of that ideal kinship model. Crucially, though, the effect of such overlapping timings become extremely visible and tangible, as it is the bodies of those men and women that end up—voluntarily or involuntarily—subscribing to, or being excluded from, that kinship ideal, that bear the different marks of that temporal binding. Behaviours, pictures, and movies all become precious sources in the research archive.

treat bodies 'as stable or pre-political' (Wilcox, 2015: 8). Applied to victory research, an attention to extratextual markers of victory can help make sense of the co-constitutive relationship between the event-like temporality of victory and its material nature that lie at the heart of hegemonic understandings of the concept. Indeed, it aligns with 'an investigation of war's empirical fields' that is attentive to 'the formal and tacit knowledges, embedded practices, bodies, affects, institutional cultures, technical machines and mundane objects through which armed conflict is actualized every day' (Bousquet, Grove, and Shah, 2020: 112).

The broader theoretical point, here, is that a temporal apparatus for the study of victory that seeks to remain attentive to its politics cannot dispense with a complementary focus on the materialities of both war and time. I suggest doing so by extending the victory research archive beyond written and oral texts. Extratextual signifiers of victory—such as the video footage in the Bint Jbeil example already discussed, triumphal arches and other victory monuments and artefacts, as well as supposedly victorious bodies—can also conceal dynamic operations of timing. These too are timing indexicals. These markers of victory and the narratives articulated through and about them, it follows, should form an integral part of the victory research archive alongside speeches, military strategy, academic theories, and other utterances of victory. That these sources currently remain largely neglected in scholarly studies of victory—recent calls to foreground bodies and matter in the study of war (e.g., McSorley, 2013; Wilcox, 2015) notwithstanding—further underscores the importance of analysing them.

Indeed, certain bodies and artefacts explicitly signify victory and should be treated as the material parallel to victory references in oral and written texts. In turn, the temporal traces that these material symbols harbour, like scars and other signs, often work as temporal clues, offering insights into the timing operations articulated through these objects. For instance, visual clues like tattoos do not only support the emplotment of war narratives, but also acquire distinctive value on account of their embodied nature (Chapter 4). Similarly, the materiality of victory monuments, for instance, lands credibility to mythical identity narratives by pinning them down spatially (Chapter 5).

The material, and more precisely, extratextual nature of these traces has specific effects on the temporal operations that they conceal. For example, the materials employed in the construction of victory monuments are themselves telling of the envisaged durability of the artefact (see Neumann, 2018: 332). The choice of medium anticipates the wilful political and social implications to which it gives rise, shedding light on the commemorative politics both of the artefact itself, and those linked to the 'will to time' of its author(s). The use of long-lasting construction materials, as opposed to perishable ones, for instance, normally signifies a desire to extend the time frame of a given victory or offer a durable reminder of it. Similarly, the inscription of conflicts' starting and ending dates lends credibility to event-like temporality of military victory and works towards the reification of wartime as exceptional and temporary. In short, these symbols still *time*, but do so extratextually.

As in the case of the queer and feminist scholarship above, however, my call for the analysis of extratextual temporal sources is not just aimed at addressing a wider, or more accurate, archive. I am interested in the complementary insights that come to the surface on account of the different modes of timing that these might conceal. And just like these contributions, I want to do so without introducing, or reproducing, a harsh separation between language and materiality (see Wilcox, 2015).

I therefore suggest grasping these different temporal dynamics through a heuristic distinction between victory narratives and victory symbols. The timing operations that the idea of victory conceals frequently take the form of victory narratives: oral, textual, or even visual accounts of victory that, either explicitly or implicitly, articulate a specific configuration of victory's temporal parameters. A historical account of a specific victory, the typologies of victory discussed earlier, and political speeches rehashing victories of the past are common examples. Other times, victory-related timing operations take the form of victory symbols: textual and extratextual signifiers of victory that hide and naturalize victory's temporal parameters. A victory arch, a trophy, as well uncontextualized victory utterances and exhortations in political speeches are all good examples of victory symbols.

This distinction between victory narratives and symbols, then, does not coincide with that between linguistic and extralinguistic signifiers of victory. It is also not a rigid distinction, but rather a tool, or expedient, that helps clarify the various roles that victory can play as a timing device. Victory narratives, for example, are more likely to be taken up as timing standards than victory symbols—e.g., detailed historical accounts of specific battles as crucial to the end of a given war. From a temporal standpoint, however, the difference between them is comparable to that between time (symbol) and timing (narrative): the latter appears to be more intuitively dynamic than the former, but this is just an illusion. Victory narratives and victory symbols, in other words, are different in *form*, but not in *nature*.

Theoretically and methodologically, then, both narratives and symbols of victory behave similarly: they consistently work as ordering tools. 'Narratives', writes Annick Wibben (2011: 2), 'are sites of the exercise of power;

through narratives, we not only investigate but also invent an order for the world'. Indeed, she further specifies, '[t]he framing of events in a particular narrative always has implications for action because it includes and excludes options and actors, while also limiting what can be thought or said, thus eventually imposing silences' (Wibben, 2011: 3). Victory narratives openly serve this function, in the sense that we can attend to them and explicitly identify events, actions, actors, and silences. The same is true for victory symbols, such as a single victory utterance or a monument: they also frame, synthesize, and order, just less straightforwardly and often implicitly or silently. As a result, the victory practices under analysis in the remainder of the book bring together both narratives and symbols of victory.

## Conclusion

This chapter set up a conceptual framework for the temporal study of victory. Building on recent work in IR on time and temporality and the critical study of war, I suggested homing in on the link between victory and time by treating references to victory as timing indexicals. Not only does this approach offer a systematic method for the examination of textual *and* extratextual victory references, but it also enables both narrational and metanarrational analysis. From a practical standpoint, this entails pausing after every encounter with a victory symbol or narrative to enquire into the ordering operations that underlie those references: which processes do they bring into focus? Why? How? By whom? Posing these questions can help identify the exact temporal relations that victory references conceal. It also sheds some valuable light on the politics of these processes. To fully grasp these temporal mechanisms and their sociopolitical implications, I pointed out, the victory research archive must be expanded. Indeed, hegemonic understandings of victory do not only posit that victory *ends* war, but also that it is material in nature: it is determined by military facts. The latter idea is not ancillary to, or separate from, the link between victory and time, especially the end of war. This material dimension of victory also affects its temporal politics. In line with the outward-looking, scalar set-up of the book, the next chapter analyses contemporary military doctrine in the UK as an exemplary victory practice, and focuses precisely on this relationship between the narrational and material dimensions of military victory.

2

# It's Time to Tell We Won

## Military Victory and the Temporal Politics of British Military Doctrine

### Introduction

Doctrine writing is an excellent example of a very intuitive, yet seldom explored, victory practice. As a corpus of knowledge explicitly concerned with collecting fighting instructions (broadly conceived), military doctrine discusses military success and victory directly and extensively. On the other hand, the narrow focus on warfighting often results in highly technical texts specifically designed to guide and instruct military personnel—and therefore rarely of interest for scholars of victory. Just like military strategy and theory of the past, indeed, contemporary doctrinal thought focuses on 'how' to win wars rather than exploring what victory means in the first place (Bartholomees, 2008: 25; cf. Martel, 2011a: 56). Nevertheless, doctrinarians do articulate victory narratives or at least employ the term as symbol, thereby configuring a specific relationship between victory and temporality either explicitly or implicitly.

This chapter analyses this victory practice by looking at contemporary military doctrine in the UK. It focuses on how doctrinarians imagine victory and its link with time. While I find that the latest British doctrine does subscribe to hegemonic understandings of victory as a time-altering event, I also find it postulates that military outcomes are the product of both 'facts on the ground' and various narratives about them. More specifically, UK doctrinarians make sense of victory through what I call a 'nested model of success', in which individual, military contributions to a conflict work as 'building blocks' of an overarching victory—but effective narratives about this victory and its building blocks can be as powerful as to radically change them and their temporal qualities.

Through the analysis of doctrine, the chapter offers a first insight about the relationship between victory and time: even military establishments reject a

*Military Victory Beyond the Battlefield*. Mirko Palestrino, Oxford University Press. © Mirko Palestrino (2025).
DOI: 10.1093/9780198955818.003.0003

strict equation between the end of war and fighting. In UK doctrine, victory is not the outcome of fighting, but of knowledge production processes that are both narrational *and* material. I argue that victory's material (or military) and narrational (or political) dimensions should be theorized as co-constitutive rather than discrete. I also highlight that failing to account for this relationship has important political consequences. For example, narratives can also be violent. While illusory, the promise of final victories that are wholly military in nature can entice a wide range of audiences. Narratives can also have fundamental effects on the materialities of warfare—e.g., by dictating how operations are planned and carried out.

I focus on UK doctrine for two key reasons. First, while the definition of military doctrine is disputed (see, for instance, Johnston, 2000; Daddow, 2002: 157; Palazzo, 2008: 6; and Høiback, 2011: 880), it is still possible to clearly identify officially sanctioned doctrinal sources in the UK thanks to the adoption of an institutionally endorsed doctrinal architecture—there is, in other words, a neatly defined set of texts that codifies the doctrine. Second, most UK doctrinal sources are publicly available on the website of the Ministry of Defence and older texts can be consulted in dedicated archives in and around London—i.e., UK doctrine is arguably among the most practical and accessible.[1]

## Victory in contemporary British doctrine: a 'nested model of success'

The publication, standardization, and organization of doctrinal texts in the UK is currently the responsibility of Defence Futures,[2] previously known as the Development, Concepts and Doctrine Centre (DCDC) or 'the Ministry of Defence's independent think tank' (MOD, 2016: cover).[3] The DCDC adopts a heuristic categorization of UK doctrinal sources into higher-level

[1] While the arguments in this chapter are specific to the UK military, they are likely to hold true for other NATO countries too. Indeed, NATO members are required to align and integrate their national doctrine with that of other member states for operational reasons. Most importantly, insights from this chapter do map onto the hegemonic understanding of victory that I find to be at work at increasingly wider geographical levels in subsequent chapters.

[2] This new organizational set-up was announced on 1 July 2024 as part of 'Programme SOLARIUM—Strategic Command's transformation programme (see MOD, n.d.). While Defence Futures is now part of the Integration Design Authority, the work of the DCDC considered in this analysis remains unvaried—at least for the time being.

[3] As elsewhere in the West, the production of contemporary, codified (see Palazzo, 2008) military doctrine in the UK spiked significantly in the aftermath of the Cold War (Mileham, 2004a: 81), fuelled institutionally by (and through) the creation of the Directorate General of Doctrine and Training in 1991 (Mileham, 2004b: 182)—renamed Directorate General of Doctrine and Development in 1998 (see Farrell, 2008: 801, fn. 123), and later on Development, Concepts and Doctrine Centre (DCDC).

and lower-level doctrine—the former category encompassing philosophy and principles, the latter practices and procedures (British Army, 2011: 3–1). Doctrinal texts, however, are numerous, frequently of classified nature, and often subject to harmonization processes. Drawing an exact picture of the UK doctrinal architecture, therefore, is not an easy task.[4] As this chapter seeks to understand how victory is conceptualized within the UK doctrinal apparatus, it builds mainly on higher-level sources on account of their shared, vocational focus on the link between politics and warfare. Indeed, it is precisely in the process of translating theories, philosophy, and principles of war into pragmatic operational guidance for the armed forces that doctrinarians encounter the victory question. By contrast, lower-level doctrine is more narrowly focused on offering pragmatic guidance of a technical nature, and therefore less prone to openly confront high-level, abstract notions such as victory.[5] Accordingly, the doctrinal sources considered in the analysis below are mainly of three types: Joint Doctrine Publications (JDPs), service-specific high-level doctrine, and Joint Doctrine Notes (JDNs).

Authored by the DCDC, Joint Doctrine Publications are the highest-level sources within the UK doctrinal apparatus and are further organized into two categories: capstone and keystone doctrine. At the top of this hierarchy is JDP 0-01, *UK Defence Doctrine* (DCDC, 2014a), the UK joint capstone publication articulating the 'broad philosophy and principles underpinning how Defence is employed' and providing 'the basis from which all other subordinate national doctrine is derived' (DCDC, 2014a: v). Right below in the hierarchy of JDPs are three types of keystone doctrine: thematic, environmental, and supporting doctrine.[6] Taken together, these doctrinal sources

[4] The latest version of the *Developing Joint Doctrine Handbook* (DCDC, 2013)—withdrawn on 28 January 2020 as a new version is in the making—for instance, advances a description of the UK doctrinal architecture that diverges from the present one on at least two counts. Firstly, it identifies three, rather than two, 'levels of doctrine', adding to high-level (strategic) and low-level (tactical) doctrine, the intermediate theatre-level (operational) doctrine—albeit acknowledging that 'there can be overlap' among them (DCDC, 2013: 1–6, 116). Secondly, it further categorizes doctrine into functional, thematic, and environmental (DCDC, 2013: 1–7)—rather than thematic, environmental, and *supporting*, as suggested below. On yet another slightly different account, the *Army Doctrine Primer* (British Army, 2011: 3–1 to 3–5) espouses both the two-tiered and three-tiered 'levels of doctrine' models, using the latter as a specification of the former. Moreover, it suggests grouping environmental, thematic, and functional joint doctrine as 'Joint Supporting Publications', collectively subsumed to Joint Keystone Publications (see British Army, 2011: figure 3:2). The problem with these conceptualizations of the UK doctrinal architecture, however, is that they increasingly grew inconsistent with the terminology employed in most recent doctrinal publications—in which, for instance, 'keystone' is used to define doctrinal texts which the *Army Doctrine Primer* would define as supporting, functional doctrine (e.g., JDP 4-00 *Logistics*), and the inter-level subordination of sources is often unspecified.

[5] Examples of lower-level doctrine are provided in footnote 10 below. In this chapter, I revert to this type of sources only to check for consistency and corroborating claims, while focusing the analysis on high-level doctrine.

[6] At the time of writing, the officially sanctioned and publicly available list of UK joint keystone thematic doctrine includes: JDP 01 *UK Joint Operations Doctrine* (DCDC, 2014b), JDP 02 *UK Operations: the*

outline the functioning of the British military and contextualize its role within the broader state apparatus. In other words, JDPs provide insights into the functioning of all branches of the military (i.e., Army, Navy, and Air Force), specifying the wide array of political activities they might be involved in—from warfighting to stabilization or disaster relief operations—as well as their contribution to each.

On account of their general nature, these are the texts that offer more reflections on victory, but also some pragmatic guidance on how to combine different instruments to achieve success. Importantly, for a fuller picture of the UK high-level doctrine, NATO doctrinal texts should also be considered. Indeed, for the sake of increased interoperability and military efficiency, the UK 'adopt[s] NATO doctrine (Allied joint publications) rather than producing national doctrine' (MOD, n.d.). As a result, most recent JDPs would specify—usually in their 'Abstract' section—existing 'linkages' with both national and NATO doctrine, detailing which of the documents within the NATO doctrinal architecture applies for the relevant theme being discussed, and whether previous national documents have been withdrawn or still complement extant NATO doctrine.[7] Similar in nature and equally numerous, these sources are also hierarchically organized and assume a comparable structure to the one outlined above for JDPs.[8]

*Defence Contribution to Resilience* (DCDC, 2021b), JDP 04 *Understanding and Decision-making* (DCDC, 2016a), and JDP 05 *Shaping a Stable World: the Military Contribution* (DCDC, 2016b). As outlined in JDP 01, 'these publications provide a JFC [Joint Force Commander] and their staff with a coherent portfolio of national doctrine, and the context for the supporting NATO and national doctrine' (DCDC, 2014b: iv). Seeking instead to detail service-specific contribution to national power, UK joint keystone environmental doctrine include JDP 0-10 *UK Maritime Power* (DCDC, 2017a), JDP 0-20 *UK Land Power* (DCDC, 2017b), and JDP 0-30 *UK Air and Space Power* (DCDC, 2017c). In turn, all remaining JDPs can be heuristically conceived as supporting doctrine—such as JDP 2-00 *Understanding and intelligence support to joint operations* (DCDC, 2011), or JDP 3-46 *Legal Support to Joint Operations* (DCDC, 2018). Other JDPs included in this list at the time of writing are: JDP 1-05 *Personnel support for joint operations*, JDP 1-10 *Captured Persons*, JDP 3-51 *Non-Combatant Evacuation Operations*, JDP 3-52 *Disaster relief operations*, JDP 4-00 *Logistics for joint operations*, and JDP 6-00 *Communications and information systems support to joint operations*. While not explicitly identified as a JDP, *Cyber Primer* is also included in the list—available at https://www.gov.uk/government/collections/joint-doctrine-publication-jdp.

7 Far from leading to a neatly bounded, pyramid-like snapshot of abiding doctrinal sources, following these 'linkages'—where secrecy allows to do as much—reveals that an intricate, web-like representation of the UK doctrinal architecture is arguably a more accurate portrait. For the sake of brevity and clarity, the analysis that follows only considers NATO's highest-level doctrine publication series, the Allied Joint Publications (AJP), insofar as it integrates extant endorsed military doctrine in the UK.

8 At the top of the hierarchy is AJP-01 *Allied Joint Doctrine*, 'the capstone NATO doctrine from which all other doctrine is derived' (NATO, 2017: viii). Keystone doctrine publications follow immediately, in the form of: AJP-2 *Allied Joint Doctrine for Intelligence, Counter-Intelligence and Security* (currently unavailable), AJP-3 *Allied Joint Doctrine for the Conduct of Operations*, AJP-4 *Allied Joint Doctrine for Logistics*, AJP-5 *Allied Joint Doctrine for the Planning of Operations*, and AJP-6 *Allied Joint Doctrine for Communication and Information Systems*. In this chapter, these are the only texts taken into consideration—when and where relevant—in the investigation of the UK doctrinal imaginary. The interested reader can enter the maze of the NATO doctrinal apparatus from either the UK MOD dedicated website or the NATO online archives.

The second type of UK doctrinal sources providing insights on military victory, then, is service-specific, high-level doctrine. Not unlike joint keystone doctrine, these sources are found right below the UK capstone doctrine and seek to elaborate on the contribution and functioning of each military branch—i.e., the British Army, the Royal Navy, and the Royal Air Force. However, service-specific doctrine differs from JDPs in terms of authorship and the adopted perspective. Indeed, these texts are developed within each service and emphasize the general principles and philosophies guiding land, maritime, and air forces, to integrate and operationalize centrally developed doctrine (i.e., from DCDC or NATO)—be it national or allied joint doctrine. The three key texts here are *Army Doctrine Publication Land Operations* (British Army, 2017), the *Royal Navy's Book of Reference 4487: The Fighting Instructions*,[9] and *AP 3000: British Air and Space Power Doctrine* (RAF, 2009b). The divergence in institutional authorship notwithstanding, all three texts highlight that they are firmly grounded within the broader, national doctrinal architecture and that they need to be read and approached as such—that is, they still refer back to, and develop, DCDC- and NATO-authored doctrine, at all levels.[10] When it comes to victory, then, these texts expand on the generic guidelines introduced by JDPs, outlining instead how armed forces in each service should be deployed—i.e., according to which specific strategies and principles—to maximize success.

The last, and more disputed, category of residual higher-level doctrinal sources is known as Joint Doctrine Notes (JDNs). The DCDC official webpage specifies that JDNs are 'raised to either encourage debate, place "markers

[9] To date, this text remains classified until further notice (see https://discovery.nationalarchives.gov.uk/details/r/C2054099)—hence could not be consulted for the purpose of this analysis.

[10] Indeed, if we were to descend into lower-level doctrine, an even more complex, heterogeneous, and intricate architectural set-up would arise. Lower-level doctrinal sources are not only more numerous than higher-level ones; authorship is even more varied, as is their content (see British Army, 2011: 3–5; and 2017: 3–5 for vivid—and yet partial—graphic representations). For the sake of both expediency and simplicity, I suggest conceiving lower-level UK doctrine as some sort of extension of the higher-level hierarchy. Accordingly, all types of sources introduced above are at least paralleled in lower-level doctrine. For instance, national joint tactical doctrine exists in the form of the Joint, tactics, technique and procedures (JTTP) series—the only publicly available of which, to date, is JTTP 4-05 on operational infrastructure. Similarly, Allied Tactical Publications provide for alliance-wide (i.e., NATO) lower-level (i.e., tactical) doctrine. Where the breadth of lower-level doctrine really starts expanding exponentially, however, is at the single-service level. For the UK Army alone, for instance, ADP Operations' graphic representation of lower-level doctrinal sources (British Army, 2017: 3–5) suggests they include: Army Field Manuals (AFMs)—further divided into three categories, namely thematic, environmental, and functional doctrine—Handbooks, Procedures, Aides Memoirs, and 'Special to Arm Doctrine'. As the level of secrecy increases the lower we descend into the UK doctrinal architecture, ascertaining the magnitude of these sources becomes harder. Nevertheless, archival records at the Liddle Hart Centre for Military Archives, King's College London from September 2000 suggests that, at that point in time, this army-specific body of tactical doctrine encompassed over thirty texts—AFM Vol. 1, in its nine parts; AFM Vol. 2, in its five parts, four handbooks, tactical aide-memoirs in three parts, special to arm inserts, and a seven-parts manual on Nuclear, Biological and Chemical (NBC) defence (British Army, 2002: 101-A-1).

in the sand" or capture and disseminate best practice', further underlining that '[t]hey are not endorsed national doctrine' (MOD, no date). Similarly, the *Developing Joint Doctrine Handbook* (DCDC, 2013: 1–5, 114) classifies JDNs as 'unendorsed doctrine', clarifying that their purpose is to quickly fill in a 'doctrinal gap' via the advantage of a less 'rigorous staffing proces[s]'. Unsurprisingly, then, the specific content of JDNs varies significantly. Two of the latest JDNs, for instance, tackle such diverse themes as *Air Manoeuvre* (JDN 1/20), and *Threat Finance and the Economic Levers of Power* (JDN 2/20).[11] Their nature as 'unendorsed doctrine' notwithstanding, JDNs offer insights into the 'ongoing' military thinking at the DCDC—a sort of 'real time snapshot' of undergoing doctrinal developments. For the purposes of this chapter, then, JDNs aptly integrate JDPs and service-specific doctrinal content by giving access to any state-of-the-art reflections on victory currently happening at the DCDC.

Working with this array of sources, the remainder of this section shows that military doctrine provides important insights into the notion of victory. In particular, I argue that contemporary doctrinal thought in the UK makes sense of victory through the articulation of a 'nested model of success'. While admittedly concerned with the question of how to achieve victory in warfare, the model simultaneously postulates some key theoretical assumptions on what victory is and proves particularly useful when it comes to illuminating the relationship between narratives and military developments or 'facts on the ground'. Notably, as a direct account of victory and its functioning, this 'nested model of success'—and British doctrine in general—should be understood as a victory practice in its own right and analysed as such.

## A 'nested model of success'

Opening the very first chapter of the UK highest doctrinal source, the *UK Defence Doctrine* (DCDC, 2014a: 2, emphasis in original), is the following quote from Sun Tzu's *The Art of War*: '[t]hus it is that in war the *victorious strategist* only seeks battle after the victory is won, whereas he who is *destined to defeat first fights* and afterwards looks for victory'. After all, the text seeks to officially 'outlin[e] the broad philosophy and principles underpinning how Defence is employed' (DCDC, 2014a: v), homing in on the policy–strategy

[11] JDNs are available at https://www.gov.uk/government/collections/joint-doctrine-note-jdn. Since writing this analysis in 2022, a new note has been published on *Intelligence, Surveillance and Reconnaissance* (JDN 1/23).

nexus, the features of the 'military instrument', and its utility. Nevertheless, the only direct references to victory in the document are those in Sun Tzu's quote above, an exhortation to the UK armed forces that '[w]hen required, we must be able to fight and *win*' (DCDC, 2014a: 25), and an assertion that said armed forces are only credible—i.e., capable of delivering an efficient 'threat of force'—if 'battle-*winning*' (DCDC, 2014a: 64).

Indeed, the *UK Defence Doctrine* emphasizes three key points. First, it specifies that the military is only one of three instruments of national power along with the diplomatic and the economic instruments—all three 'underpinned by information' (DCDC, 2014a: 12). Second, that 'military strategy is a subset of Defence strategy'—which in turn is part of a 'national strategy' articulated by the government in line with its national policy—that 'directs the military instrument' (DCDC, 2014a: 10). Third, that the political nature of strategic objectives forecloses the possibility of 'an exclusively military solution to conflict' (DCDC, 2014a: 13). In short, 'the military instrument ... is "not an independent phenomenon, but the continuation of policy by different means"' (DCDC, 2014a: 13). Nevertheless, the text provides some guidance for, and context to, 'success' in warfare. In particular, a paragraph about 'certain enduring features' of war suggests that '[t]he measure of [war's] success is the extent to which belligerents deem its political outcome to be favourable', and that both military and non-military means might be employed 'to seek success' (DCDC, 2014a: 18). Moreover, a classic distinction between strategic, operational, and tactical levels of warfare is introduced. The acknowledgement of inter-level overlap and partial blurring notwithstanding, 'strategic success' is said to hinge upon 'a combination of military force, diplomacy and economic measures', and tactical activities found to 'contribute to success on operations' as well as 'setting the conditions for success' (DCDC, 2014a: 20–21).[12] While other sparse references to success are found throughout the document, no clearer indications for its achievement—or its meaning—are provided. Instead, the 'quality of [the UK] fighting power' in general—and 'relative superiority' in particular—is generally identified as crucial to success (DCDC, 2014a: 26–27); 'high morale', 'surprise', and 'concentration of force' (DCDC, 2014a: 30–31), as well as 'confident leadership', a 'warfighting ethos' (DCDC, 2014a: 36), and a 'single, unambiguous aim' (DCDC, 2014a: 50) are also recognized as key variables in the success equation.

[12] This chapter of the *UK Defence Strategy* does not introduce the notion of 'operational success'. However, the notion is taken up in chapter 2, where among the 'UK principles of war', 'maintenance of morale' is identified as 'crucial for operational success' (DCDC, 2014a: 30).

By contrast, Joint Doctrine Publication 01: *UK Joint Operations Doctrine* (DCDC, 2014b) seems to articulate a much more specific framework for 'success'. Interestingly, the document makes no references at all to the notion of victory—a key insight to which I return below. Nevertheless, a 'nested model of success' starts to emerge as the text zeroes in on the planning, conduct and sustainment of military operations—after all, *UK Joint Operations* explicitly intends to 'assist a JFC [Joint Force Commander] and their staff on operations and in their education and training' (DCDC, 2014b: iii). Having located defence governance within the broader national political landscape—reinforcing the fundamentally political nature of both national and defence strategies teased out in detail in the *UK Defence Doctrine* above—the document delves into the procedural aspects of 'military planning and decision-making' (DCDC, 2014b: 12–14). The ultimate responsibility to 'identif[y] the military conditions for success'—that is, at the highest, or most general level—rests with the Chief of the Defence Staff (DCDC, 2014b: 12). Importantly, such conditions are to be consistent with (and considered for analysis within) a 'national strategic aim'—that is a government-defined 'future desired outcome' to be achieved via a combination of all three instruments of power—and the (slightly) 'narrower' 'national strategic objectives' designed to lead to its attainment (DCDC, 2014b: 16). Within this apparatus, the 'military strategic end-state'—in turn to be achieved via 'military strategic objectives'—delineates 'the extent of the military contribution', specifying the commitment of armed forces and their alignment with the other, non-military instruments of power (DCDC, 2014b: 17).[13] As *UK Joint Operations* specifies, '[a]n objective is a clearly defined and attainable goal for a military operation, for example seizing a terrain feature, neutralizing an adversary's force or capability or achieving some other desired outcome that is essential to a commander's plan and towards which the operations is directed' (DCDC, 2014b: 19). Indeed, the hierarchy of aims and objectives described so far (see DCDC, 2014b: 19, 25 for a summary) is nothing but the tip of a much larger iceberg.

The attainment of a given 'military strategic objective' requires different Joint Force Commanders to meet their 'operations objectives'—that is, specific objectives within 'their own', 'single theatre of operations' (DCDC, 2014b: 20). To that end, following a concerted process of analysis (see DCDC,

[13] While largely overlapping, the NATO structure for multinational operations adopts a slightly different terminology (see NSA AAP-06, 2013; NATO, 2017). It suffices to stress here, for the purpose of the chapter, that the NATO model shows a similar 'nestedness'.

2014b: 39), each Joint Force Commander[14] defines military actions and tasks through which 'strategic ends' are linked 'with tactical action' (DCDC, 2014b: 40; see also 87–92). In turn, commanders often allot responsibilities for task(s) fulfilment down the command chain, and 'subordinates remain accountable to their superior for executing assigned tasks' (DCDC, 2014b: 101). Nevertheless, the doctrinal text suggests that '[c]ommanders have sole responsibility for ensuring that their plan delivers the best chances of success' (ibid.), adding that 'commanders' *success* depends largely on *experience, flexibility* and *determination*' (DCDC, 2014b: 102, emphasis in original). I return to this explicit association between success and (in this case a commander's) personal propensities and traits—expressed also in terms of 'a combination of ability, intellect and intuition', 'instinct and vision', 'talent', or 'irrational tenth' (DCDC, 2014b: 105)—below. Here, it suffices to emphasize that, the attribution of responsibility to commanders notwithstanding, the nested nature of the model presented so far is further reiterated and expanded upon as we 'descend' into lower levels of war.

Within the Royal Air Force's single service tactical doctrine, for instance, the continuation of a nested model of success is encapsulated by a doctrine-mandated process known as 'strategy-to-task' (RAF, 2009a: 1–5, 3–1 to 3–10). Designed to 'demonstrate that every target attack contributes towards the success of a campaign' (RAF, 2009a: 3–2), strategy-to-task is an 'effects-based approach ... that focuses on the end-state' by breaking it down into 'friendly goals' in the form of—from the top, down—'operational air objectives', 'operational air sub-objectives', 'tactical air objectives', 'tactical air sub-objectives', and, finally, 'tactical air tasks' (RAF, 2009a: 3–2, 3–3). The approach is based on the assumption that '[t]he strategic aim of a campaign is always political' (RAF, 2009a: 3–1) and that good planning should start from that 'desired end-state' and translate it into narrower, more manageable objectives and tasks—each feeding into the immediately 'higher' level if approached in the reverse sense. Within the UK doctrine, in other words, success is imagined as the product of 'small' building blocks, each of which is allotted to different echelons at different levels, with the aim of contributing to an 'overarching' success resulting from their sum.

[14] Added layers of complexity here are non-military actions contributing to a military objective—for which, non-military entities and personnel might be involved—as well as the outsourcing of certain military activities to private military companies, and operation design and planning conducted at the multinational level (for an overview see DCDC, 2014b: 43–51).

Importantly, a few additional fundamental provisos are found within the doctrinal imaginary that sensibly complicate this, otherwise fairly linear and straightforward, nested model of success. Indeed, while translating and extending the model to Army-specific operational and tactical doctrine—especially through 'Integrated Action' and the 'Manoeuvrist Approach'—*Army Doctrine Publication Land Operations* (British Army, 2017: P1) simultaneously warns that 'tactical success is increasingly difficult to achieve in purely physical terms' and that 'it is not possible to translate tactical success into desirable political outcomes without gaining favourable consensus among multiple audiences'. The first specification can be understood as a reflection of the complexities of defence strategy as defined in the higher-level, joint doctrinal texts discussed above—both in terms of the non-exclusively military nature of overarching strategic objectives, and the combination of lethal and non-lethal means which might be required for the accomplishment of any lower-level military action.

The second, however, introduces a more radical idea on the centrality of narratives which cuts transversally through the whole model. The starkest and clearest expression of this tenet is found in both the forewords to *Army Doctrine Publication Land Operations* (British Army, 2017: i) and JDP 0-20 *UK Land Power* (DCDC, 2017b: iii) in the following formulation: '[s]uccess is more likely to be achieved through non-military or non-lethal means, and *invariably it is the triumph of the narrative that is decisive*, not necessarily the facts on the ground' (my emphasis). Elsewhere in the text, *Land Operations* articulates the idea in more details:

> As the audience's judgment is an increasingly significant factor in contemporary conflict, operations must be designed and conducted accordingly. No conflict has a purely military solution, and overall success requires favourable consensus among a diverse audience. How we say, how we behave and what we do, influence how we are seen must be consistent and appropriate [*sic*]. *At the margin, a neutral or nearly neutral outcome of military action can be turned into a success or a failure by how it is perceived.*
>
> **(British Army, 2017: 1–9, my emphasis)**

In other words, a premium is put on perception, and therefore on narratives, to the detriment of facts on the ground—that is, of the material 'outcome of military action'. Far from being Army-specific, the idea finds resonance with similar positions within the broader doctrinal imaginary. The *UK Defence*

*Doctrine*, for instance, suggests that 'a strategic narrative ... is crucial to efficiently managing information, as well as *influencing a range of audiences and activities* in a consistent, coherent manner' (DCDC, 2014a: 15, my emphasis).[15] Coming closer to the Army doctrine here above, JDP 0-10 *UK Maritime Power*, acknowledges that '[w]hen attempting to generate or project influence it is likely that *the prevailing narrative and not necessarily the* facts that will prove decisive in determining success or failure' (DCDC, 2017a: 18). More recent developments in the UK doctrinal thought elaborate on the idea even further, highlighting its continued and vivid relevance. In particular, Joint Doctrine Note 2/19 *Defence Strategic Communication* redefines success altogether, stating bluntly that '[s]uccess is achieving the behaviours (changed or maintained) by target audiences that the UK government intended' (DCDC, 2019: 19; see also 3) and that '[m]ilitary actions and tactical victories are only important insofar as they contribute to such behaviour changes' (DCDC, 2019: 23). Accordingly, a narrative is what makes the cut: '[a] narrative is generally defined as: a spoken or written account of connected events; a story. ... A target audience's "truth" is what it perceives to be true (irrespective of the facts) and its belief in that truth can be hard to shake, even when presented evidence of the facts' (DCDC, 2019: 6).

On a similar note, one recent doctrinal publication, *Integrated Operating Concept* (DCDC, 2021a: 6), reiterates the idea and seeks to consolidate it further, suggesting that '[t]he triumph of the narrative increasingly determines defeat or victory and hence the importance of information operations'. The text suggests that 'Russia's seizure of Crimea in 2014' is a clear example of this dynamic: 'a stark case study in which a fait accompli strategy changed facts on the ground below the threshold at which a warfighting response would be triggered' (ibid.). Indeed, the Russian success in seizing Crimea, *Integrated Operating Concept* explains, rested on a skilfully crafted and aptly conveyed narrative that enabled deploying 'specialist military forces' in Ukraine while simultaneously denying that intervention, convincing varied audiences that 'they were ethnic-Russian nationalist militia' (ibid.). In short, success was achieved via a narrative that was able to change facts.[16]

[15] JDP 01 *UK Joint Operations Doctrine* adopts a similar position (see DCDC, 2014b: 16, 20) and provides some details as to the need of the narrative to efficiently integrate both military and non-military contributions under its umbrella.

[16] To quote *Integrated Operating Concept* once more, '[u]ltimately, this activity affected the perceptions and beliefs of the local population, Ukrainian leadership and the international community, allowing strategic objectives to be swiftly achieved following the deployment of conventional military forces' (DCDC, 2021a: 6).

## A false dichotomy: victory's material and narrational components

Not unlike earlier military thinkers and strategists, contemporary doctrinarians seem more concerned with the question of 'how to achieve victory' than with clarifying the concept's meaning (Martel, 2011a, 2011b). As a result, contemporary doctrinal sources also work with a notion of victory that continues to be murky and difficult to grasp (see Mandel, 2006, 2007; Bartholomees, 2008; Martel, 2011a, 2011b; Hom, O'Driscoll, and Mills, 2017; O'Driscoll, 2020a). Nevertheless, in the process of articulating the nested model of success sketched out above, important insights into the conceptual and temporal parameters of the notion are given away. Firstly, success and victory designate, in line with the widest accepted position in War Studies, one of the potential *outcomes* of armed conflicts (Mandel, 2007; Martel, 2011a; Krause, 2019). Here, the influence of academic typologies of victory differentiating between levels of war (see Gray, 2002; Mandel, 2006, 2007; Zaidi, 2009; Martel, 2011a, 2011b; Krause, 2019) is also visible—at least in the use of the qualifiers tactical and strategic to specify tasks, objectives, and successes. Secondly, and relatedly, ideas of finality or temporal closure (see Hom, O'Driscoll, and Mills, 2017: 236–237; Campbell and Steele, 2021) are kept within victory's conceptual coordinates, in a notion that still designates a time-altering event separating wartime from peacetime—an 'end-state', as expressed throughout doctrinal sources. Thirdly, the political and subjective nature of victory—i.e., victory means different things for different people (see Bartholomees, 2008; Martel, 2011b)—is repeatedly reinstated, both through the emphasis of war as 'the continuation of policy' (DCDC, 2014a: 13) and the definition of success as the achievement of desired changes in one's audience's behaviours (see DCDC, 2019: 19).

However, emerging from the analysis of UK contemporary doctrine above is also, at least, a key discontinuity—or tension—with extant academic conceptualizations of victory: a widely accepted separation between the political and military outcomes of armed conflicts is simultaneously deployed and cast into doubt, or disputed. This can be grasped, I suggest, in terms of a false dichotomy between victory's narrational and material dimensions. Rather than merely subsuming the military outcomes of conflict—what doctrinal sources refer to as 'facts on the ground'—to politics, or ascribing different types of victories to different temporal phases (Mandel, 2006, 2007), doctrinarians seem to postulate a relationship between the materialities of warfare and their (re)narration that can be understood as one of co-constitution.

To take this assumption seriously, however, is to acknowledge that victory is not the outcome of armed conflicts, but of concurrent knowledge production (and timing) processes encompassing both warfare and other non-warlike—but not necessarily non-, or even less-violent—practices such as memorialization activities, memory-making exercises, and story-telling enterprises.

## Narrational/material outcomes

Dwelling between a vehemently political notion of victory and the necessity to articulate a clear model to guide armed forces to success, doctrinal sources introduce the two above-mentioned provisions on the importance of narratives. In particular, after acknowledging that 'the military instrument ... is "not an independent phenomenon, but the continuation of policy by different means"' (DCDC, 2014a: 13), doctrinal texts articulate a nested model of success within which both military and non-military operations are crucial to both the 'national strategic aim' and the 'military strategic end-state' (DCDC, 2014b: 16)—i.e., military and non-military outcomes. This is reflected by two central ideas. Firstly, non-strictly military activities such as information operations are included in the model among the building blocks that collectively lead armed conflicts to success (DCDC, 2021a: 6). Secondly, the acknowledgement is made that 'the prevailing narrative and not necessarily the facts ... will prove decisive in determining success or failure' (DCDC, 2017a: 18). Importantly, the latter proposition should not be read as restating the former. Taken in conjunction with the intuition that an 'audience's "truth" is what it perceives to be true (irrespective of the facts)', it submits, instead, that an interaction exists between the materialities of military operations and the narrational effort to report those 'facts' to one's audience(s). Not only might 'facts' change or orient the 'narrative', but 'narrative' might also change the 'facts'. In short, victory's narrational and material dimensions are co-produced.

Additional provisions found throughout doctrinal texts corroborate this interpretation. JDN 2/19 *Defence Strategic Communication*, for instance, specifies that '[s]trategic narratives must be fought', further explaining that '[d]efence's actions, images and words must consistently align with the relevant strategic narrative to build and maintain credibility. Maintaining the initiative will require *a proactive and innovative approach*. Hard-earned credibility with audiences must be protected' (DCDC, 2019: 6, my emphasis). Interestingly, the Joint Doctrine Note clarifies that '[w]here media exposure is essential to raising the profile of the messages that Defence wishes to send, *events may be planned to fit the media timetable*' (DCDC, 2019: 38, my emphasis). Indeed, '[j]ust as the narrative needs to be fought, our media messages must compete with other news stories prevalent at the time. To help generate publicity, IPS [International Policy and Strategy] teams need to consider how their activity can be made newsworthy and direct subordinate commands accordingly' (DCDC, 2019: 38–39). In other words, the story one seeks to tell about military operations is so important that the planning of said activities has to take 'the media timetable' into consideration.

Far from being an isolated claim in a yet-to-be endorsed doctrinal text, this position is consistent with doctrine of the highest level. In detailing the responsibilities of national contingent commanders, for instance, JDP 01 *UK Joint Operations Doctrine* lists '[c]oordinating and shaping national media' among their key roles—together with 'exercising operational control of national forces' or 'determining and coordinating national force protection policy and posture' (DCDC, 2014b: 113). After noticing that '[s]upportive media coverage plays a major role in maintaining the endorsement of both the international community and domestic audiences' (DCDC, 2014b: 112), the text attributes media responsibility to commanders, providing them with as pragmatic instructions as 'be[ing] honest about their ability to deal with the media and carefully balanc[ing] using a media spokesperson with their own appearances', adding that '[t]here are likely to be moments where the message is so important that they must deliver it personally' (DCDC, 2014b: 115).

An understanding of victory as co-constituted by material, military developments *and* the narratives articulated about them sits uncomfortably with extant theorizations of victory as an outcome of armed conflicts. The few existing studies of military victory that focus on defining the concept reiterate—either explicitly or implicitly—an underlying understanding of victory as a war outcome. For Robert Mandel (2006: 5–9), for instance, extant conceptualizations of victory can be grouped into 'end-state' and 'cost–benefit' types of definitions. While the former measure victory on account of the alignment between one's aims and one's results, the latter emphasize whether a given actor has 'spent' too much—or 'whether war generally has been worth the effort, justifying the lives lost and money spent' (Mandel, 2006: 7). In both cases, a balance must be struck between competing factors—and a positive balance generates a positive, or 'successful' outcome. While William C. Martel (2011a: 9; emphasis in original) differentiates between 'three specific levels of victory ... *tactical victory, strategic victory*, and *grand strategic victory*', he similarly defines victory as an outcome, only one that yields increasingly significant political effects depending on the level one considers. As he put it, his '*typology of victory* is designed to categorize it as a function of the outcome or effect of war' (Martel, 2011a: 32, emphasis in original).[17]

While evoking the 'outcome' connotation of victory that doctrinal sources reach for in their use of the concept, these accounts effectively separate the *material* outcome of war from the (material *and* narrational) process leading to it logically, temporally, and politically—or to reiterate the vocabulary of UK doctrinal sources, a fictional distinction is introduced between 'narrative' and 'facts on the ground'. Mandel's solution to the definitional problem, for example, is to further distinguish between the 'war-winning' and 'peace-winning' phases of an armed conflict (Mandel, 2006: 13–15). The former 'occurs when a state attempts to bring a war to a successful military conclusion' (Mandel, 2006: 13), and 'entails the complete withdrawal or retreat, if not laying down arms or surrender, of enemy troops, thwarting their objectives while accomplishing one's own' (Mandel, 2006: 16). The latter, instead, 'occurs when a state attempts to reap the payoffs of war' (Mandel, 2006: 13) and designates a weighted outcome over six, interrelated elements: informational, military, political, economic, social, and diplomatic (Mandel, 2006: 16). On this account, the material and political outcomes of conflicts are *explicitly* different. Albeit interrelated, military victory is 'easier to determine' (ibid.) and immediately manifest, whereas strategic victory is properly political, infringes upon other domains such as diplomacy and economics, and only happens afterwards.

Notably, this process/outcome separation persists even when heuristic distinctions between additional 'levels' of victory are introduced (see, for instance, Gray, 2002; Bartholomees, 2008; Zaidi, 2009; Martel 2011a, 2011b; Krause, 2019). To admit that 'what is won on the battlefield can be lost entirely thereafter if the countries attacked are not turned into better and safer places'

[17] Importantly, Martel (2011a: 47) also admits that '[b]eyond studying victory as an outcome, it is important to acknowledge that victory represents a political and military aspiration on the part of policy-makers before any military action is taken'. In short, for him the mere promise of victory from policymakers yields important political consequences, such as the capacity to mobilize the public in support of the war effort. Unfortunately, Martel does not explore this idea in detail, suggesting instead that further research should be undertaken to explore the notion of victory as 'aspiration', and whether any 'formal linkages' can be established between victory and 'popular support for a war' (Martel, 2011a: 50). As a result, within the brief discussion that Martel (2011a: 47–50) provides on the issue at hand, victory as an aspiration continues to resemble the idea of an outcome, only one that convincingly 'realized' in potentiality—rather than in actuality—has some political consequences. While this notion of victory as an 'aspiration' starts gesturing towards the importance of narratives, it still reproduces the fictional separation between victory's material and narrational dimensions under discussion here—or, it does not perform the epistemological shift that I am calling for.

(Mandel, 2006: 13), or that tactical and operational successes 'do not necessarily change the final outcome of the war' (Martel, 2011a: 36), is not enough. While addressing the political and subjective nature of victory, these typologies still assume military (or material) and political (or narrational) outcomes as discrete. Moreover, this reasoning concomitantly leads to a (politically problematic) confinement of 'tactics' to the military realm, and 'strategy' to that of politics. Martel (2011a: 26), for example, suggests that victories at levels which are higher than the tactical (i.e., strategic and grand-strategic) are increasingly less a concern of 'military science' as they pertain to the realm of politics. Beyond failing to grasp the interaction between victory's narrational and material dimension, these accounts contribute to concealing tactics as 'technical', 'scientific' or, worse still, 'a-political', and narratives as non-warlike, or less violent—a proposition that clashes with, for instance, the idea that 'the narrative needs to be fought' (DCDC, 2019: 38).

Considered from a temporal standpoint, this discontinuity with extant theorization of victory—and its analytical and political value—emerges even more starkly. A harsh separation between military and other levels of victory, for example, shifts the analytical and temporal coordinates of the 'victory problem' by extending the timespan of the analysis. The conceptual fog surrounding victory, however, is not lifted. Now that we know that victory is a successful outcome, we need more solid theorizing of the processes from which it stems. Indeed, further research is often called for to shed light on how to 'win the peace' after having 'won the war'. Consider, moreover, the idea of finality associated with victory as outcome. Assumptions of temporal finality are so embedded in vernacular uses of victory that it suffices to utter the term to conjure up in one's audience ideas of temporal rupture, termination, and closure. However, that the link between victory and finality is so well established does not make it any less deliberate and constructed. In fact, a tight association between these ideas only endures because we keep reinforcing it. The exquisitely temporal distinction between war- and peace-winning, for example, works precisely in this direction. Indeed, not only does the idea of an outcome already evoke finality on account of its conceptual set-up—i.e., an outcome is normally the product of a process, that which comes at the end. To make sense of victories on the battlefield that do not necessarily prove time-altering or politically decisive, a distinction is introduced between war-making and peace-making processes. As a result, different outcomes can be ascribed to different processes, and finality brought back in, for both instances. It is not the case, in other words, that military victory did not bring warfare to an end—i.e., that victory did not deliver rupture or termination. Rather, the problem is reframed analytically: finality is there even in the first phase, only that phase is conceptualized as a distinct, narrower process.[18]

Within the theoretical apparatus of contemporary British doctrine, this relationship between victory and finality is reinstated via, at least, two conceptual moves. Firstly, overarching victory in the nested model of success is identified with the expression of the 'end-state'—a 'national strategic aim or *end-state*' (DCDC, 2014a: 8, my emphasis) and a 'military strategic end-state' (see for instance DCDC, 2014b: 19). A temporally fraught formulation, 'end-state' carries the double connotation of outcome—the state at the end of a process—and of an outcome characterized by finality—the conclusion of said process (i.e., it hints at a logic of linear causality). Secondly, the end state sits at the apex of the nested model of success. Here, a subtle logic of linear continuity—if not linear causality—is instantiated whereby lower-level objectives (or building blocks) collectively feed into increasingly higher-level tasks, up until reaching that *final* end state. *UK Joint Operations Doctrine* (DCDC, 2014b: 18, 52) captures this logic graphically, illustrating lower-level objectives as subordinate, narrower blocks that feed into higher level aims: military actions generate specific effects which, in turn, engender 'conditions' for the attainment of (hierarchically ordered) objectives. Moreover, the accompanying prosaic explanation reinforces the message: military objectives at different levels are always instrumental, insofar as they serve 'to meet the national strategic aim' (DCDC, 2014b: 19). Just like typologies of victory, UK doctrinal sources might contribute to the reproduction of a notion of victory as an event, endowed with rupture-like temporal qualities that bring conflict to closure.

By specifying that facts on the ground might be changed by narratives, however, UK doctrinarians configure a different relationship between victory and time. Victory, in British doctrine, can be understood to work as a timing outcome: it presents the features of a timing indexical that conceals—and thereby subtly reproduces—a (passive) timing attempt. When using victory, in other words, a dynamic operation of ordering between different phenomena is evoked which postulates that victory comes at the end of it, usually in the form of an event that brings closure. At the risk of oversimplifying, the phenomena being coordinated in these timing operations certainly include the nested tasks and objectives of relevant military operations. Importantly,

[18] Counterintuitive as it might sound, this theoretical move is widely popular, and perhaps unsurprisingly so. The idea that military victories are becoming less decisive, and that this is the result of the 'transformation of war', or 'revolutions in military affairs', is a very well-explored research area. After all, as mentioned above, that victory designates the end of conflicts is an established tenet. Accordingly, Mandel's theory itself can—and, I suggest, should—be treated as a narrative that convincingly reinforces victory's peculiar temporal parameters.

however, UK doctrinal sources emphasize the fundamentally concurrent timing of narrational efforts aimed at persuading different audiences. In fact, 'a strategic narrative ... efficiently managing information, as well as influencing a range of audiences and activities in a consistent, coherent matter' (DCDC, 2014a: 14) might even act as the timing standard for these operations. Here, equal status should be accorded to the phenomena being timed, be they material outcomes of specific, nested operations or the stories told to varied audiences with the aim of persuading them. Notably, while both the material and narrational dimensions of the concept are proffered as worthy of analysis, it is easy to lose sight of this overarching, complex picture and only focus on one component. As any other timing indexical, indeed, victory obscures these dynamic operations, providing instead a signifier of their synthesis or, indeed, their outcome.

Moreover, on this account assumptions of linear causality such as those intrinsic to the logic of the building block above are also cast into doubt—not least because of the different temporal parameters of victory narratives and facts on the ground. That tasks and objectives lead linearly to an overarching, final victory is only one of multiple possibilities—in fact, a military objective such as the victory image that the IDF hoped to capture at Bint Jbeil (see Introduction) can be construed as single-handedly decisive. Here, the idea that victory is co-constituted materially *and* narrationally is fundamentally temporal in the sense that it emerges *through* time, or more precisely timing operations. Notably, this does not mean that we cannot think of military developments and narrative as separate, but that they are produced as such through *temporal* operations of ordering.

Put differently, a focus on time can account for victory as jointly produced by war's materialities and their narration, as well as provide a formal explanation for the reproduction of a false dichotomy between these components. It also invites us to adopt a critical posture towards the ordering operations resulting in victory. Indeed, as Hom (2018a, 2020) specifies, timing operations are multiple, human-led, and always political. The analytical implications of this proposition are varied. Firstly, the politics of victory acquire central stage. As seen above, for instance, a failure to account for the role of narratives in the victory equation might result in the sanitization of narrational enterprises as non-violent, or the construction of tactics as scientific and seemingly apolitical. Does it matter that victory is imbued with finality? What are the political consequences of deploying victory as a periodizer?

Secondly, the analytical terrain within which victory is made and remade is sensibly expanded (or scaled up). This section has shown, for instance, that the process–outcome separation discussed above naturalizes and conceals the sustained—political, narrational, and dynamic—timing operations at play in constructing victory as an outcome. By centring time—or, more specifically, the temporal assumption embedded in the concept—it becomes apparent that ideas of finality, closure, or temporal rupture are imperceptibly folded into victory when conceptualized as outcome, to the detriment of the (temporal) processes leading to it. I have argued that at least two different processes are responsible for this association. First, the use of victory as a symbol, which carries among its assumptions a well-established connotation of temporal finality. Second, the active, further corroboration of this relationship via, for instance, the naming of success as end state in UK doctrine, or the distinction between war-winning and peace-winning phases in Mandel's scholarship (2006, 2007). It follows from these observations that, in the study of victory, an added layer of analysis is called for exploring the complicity of narratives *and* metanarratives to the establishment of victory's conceptual tenets. If victory is concomitantly made materially and narrationally, then it is made as much without the battlefield, in times of peace, as within the battlefield, during wartime. This fact should be reflected in the study of victory.

In the remainder of this chapter, I first perform an analysis of the conceptual politics of victory within the UK doctrinal apparatus, in line with point one, to then sketch out some theoretical reflections on what it means to take seriously point two on narratives and metanarratives—to which Chapter 3 is devoted.

## The conceptual politics of success/victory in contemporary UK doctrine

Contemporary British doctrine features a notable tension: while proffering that the relationship between victory's narrational and material dimensions is both fundamental and complex—or, as I suggested, one of co-constitution—doctrinal texts adopt the notion of victory as an outcome of armed conflicts and reproduce a long-standing association of the concept with the idea of finality. Within the UK doctrinal apparatus, two key strategies contribute to the reiteration of finality: the expression of 'end-state' as a signifier for overall success, and the postulation of said end state as the product of the accomplishment of nested objectives and tasks. What are doctrinarians after when theorizing victory in this way? What are the conceptual politics of victory and how do they play out within the UK doctrinal apparatus?

Beyond identifying the conditions of success in the change or maintenance of one's audiences' behaviours (DCDC, 2019: 19; see also 3), doctrinal texts explicitly admit their concern with influencing a wealth of different audiences. As JDP 0-20 *UK Land Power* vividly articulates:

> People exist in linguistic, cultural, social and political groups with specific identities, usually associated with particular territories. Individually or collectively they form the audiences whose perception and interpretation of events, and subsequent behaviour and decisions, determine how conflict is conducted and resolved. These audiences may include those deployed, as well as broader international audiences, including those in the conflict region. They may be hostile, friendly or neutral to the actions of land forces, although the boundaries between these groups are not fixed.
>
> **(DCDC, 2017b: 16).**

Importantly, the quote does not only reaffirm the political aim of persuading a varied audience; it also identifies some of these audiences, including: deployed armed forces, the public opinion of the countries to which the forces belong, the other party (or parties) involved in the conflict, as well as other international observers. Indications as to how to go about persuading these audiences are also provided. JDN 2/19 *Defence Strategic Communication* (DCDC, 2019: 6–7), for example, suggests that a narrative 'is a story designed to provide an emotive justification for a policy goal' and should, therefore, 'be empathetic with the emotions and motivations of the audience'. In other words, one way to convince an audience is to tap into their affective dispositions, for example, by articulating a narrative that can align with their motivations and sensibilities.

The nested model of success under analysis here is, I suggest, an instantiation of such an emotive narrative. Firstly, military doctrine is, like all knowledge, narrational and political—or, as Daddow (2002: 158, my emphasis; see also 2003) puts it, 'doctrine is more than a collection of standard operating procedures. It is a *highly political and politicised entity*, operating on all the levels of war and seeking to develop a philosophy and attitude of mind for members of the armed forces as much as it lays down sets of technical recommendations'. In short, the narratives that doctrinarians articulate and promote have an expressly political purpose.[19] Secondly, victory is an emotionally laden notion (see Mandel, 2006: 13; Campbell and Steele, 2017, 2021; O'Driscoll, 2020a: 128) and doctrine is a means to cultivate its affect.

[19] Consider, for example, that every doctrinal text in the UK actively identifies its designated audience(s) in a 'preface' section.

For Harald Høiback (2011: 881–882, 2013: 32–38), modern military doctrine originates in nineteenth-century France, in the wake of a shattering defeat against Prussia in 1870. In this context, the French army 'not only needed a new formula for victory, but also something *to energise and inflame will power* and to foster what later became known as *élan vital*' (ibid., my emphasis; see also Høiback, 2013: 34). 'Modern doctrine', for him, 'is thus a French invention with a double purpose: to explain how to win the next war and to infuse the will to do so' (Høiback, 2013: 38; see also 2011: 883). Lastly, some patterns can be identified in the use of the notions of 'victory' and 'success' throughout doctrinal texts, which would provide for a formal explanation of this proposition—i.e., the differential use of the terms victory and success in British doctrine works as a narrational strategy to tap into the affective dispositions of its varied audience.

On the one hand, the use of victory and success as synonyms is widespread in non-specialist discourse, and not unprecedented in scholarship on warfare either. In his work on military doctrine and the generation of fighting power, for instance, Geoffrey Sloan (2012: 263) suggests that '[i]n summary, doctrine, irrespective of the level at which it manifests itself, is pivotal to success in war' and clarifies that such a claim is not new by quoting Knox's idea that 'the influence of doctrine upon victory is profound' (ibid.; for a similar instance see Gooch, 1997: 8–9). On the other hand, some scholars of victory suggest that 'victory' and 'success' are conceptually distinct, emphasizing that their conflation is not without implications. Cian O'Driscoll (2020a: 7–9) for instance, warns that, at least within the theoretical set-up of the Just War Tradition, success and victory diverge sensibly. Going back to Clausewitz's *Vom Kriege*, O'Driscoll maintains that victory 'unlike other similar terms such as "success", denotes a fundamentally zero-sum outcome' (O'Driscoll, 2020a: 6). Accordingly, the shift from victory to success that takes place in Just War Theory 'obscures the antagonistic logic of warfare that victory captures', with the result of 'nudg[ing] the tone of just war inquiry onto a euphemistic register' (O'Driscoll, 2020a: 8). Via success, that is, Just War Theory somehow sanitizes war, giving the 'impression that prevailing in war has nothing to do with the brutish, zero-sum realities of defeating the other party, when in fact it is grounded in them' (ibid.). In short, on this account, it is no accident that the closest articulation of a 'theory of victory' (see Høiback, 2011: 883, 2013: 10) in UK doctrinal texts is expressed in terms of the nested model of *success*.

As emphasized above, the word 'victory' is very rarely deployed in UK doctrinal texts. Most of these references, however, are consistently of two types. Firstly, victory is repeatedly employed in either quotes from authoritative figures—such as military thinkers or practitioners—or longer historical illustrations employed to clarify key doctrinal tenets (see, for instance, DCDC, 2014a: 2; British Army, 2017: 1–2, 8–15, 8–19, 10–10).[20] Secondly, victoriousness or other victory-derived *abilities* and *capacities* are ascribed to individuals. Like commanders in the discussion above, then, army 'leaders and soldiers' have 'a will to win' (DCDC, 2017b: 29), maritime forces possess an 'enduring spirit' which gives them a 'determination to fight and win' (DCDC, 2017a: 32), and RAF echelons have 'war-winning capabilities' (DCDC, 2017c: 43). In both cases, no explanation is provided as to the meaning of victory, a concept which is instead treated as self-explanatory. By contrast, whenever indications are provided as to how to achieve victory or what constitutes it, the employed term is success.[21] The clearest example here is the nested model traced at the beginning of this chapter, in which success denotes the accomplishment of specific objectives. The nestedness of the model ensures that these objectives feed into higher level objectives, so that the completion of each of these building blocks contributes to a broader, overarching success. Broken down into smaller tasks with narrower objectives, military operations become more manageable, and the conditions for success circumscribed and easier to identify.

From this differential use of the two terms, success emerges as a managerial and technical goal that the doctrine unpacks at length and provides clear indications for. Instead, victory is aspirational: some sort of ideal version of that successful end state of which we have historical examples, and about which we have word of wisdom by authoritative military thinkers. Importantly, victory is not removed from British doctrine, but kept instead as a benchmark the achievability of which is proved by history.

One way to interpret this peculiar choice is to consider the affective potential it unleashes. According to Luke Campbell and Brent Steele (2017, 2021) victory's affective potential is tightly linked to its ambiguity. '[A]mbiguity',

[20] The text that contains most references to victory, for instance, is Army Doctrine Publication: *Land Operations* (British Army, 2017), the British Army's highest, branch-specific doctrinal source. Importantly, in all six instances, victory is deployed as self-explanatory. Victory is employed twice, for instance, to refer to the Duke of Marlborough's achievements at Blenheim in 1704 (British Army, 2017: 10–10). Similarly, the term appears in three quotes from authoritative figures, namely Mao Tse Tung, General Stonewall Jackson, and General Rupert Smith. The first introduces a section on 'the nature of conflict' and explains that 'unless you know the actual circumstances of war, its nature and its relation to other things, you will not know the laws of war, or how to direct war and how to win victory' (Mao Tse Tung, cited in British Army, 2017: 1–2). The remaining two quotes corroborate a chapter on 'orchestrating and executing operations' but are similarly vague in content—one suggesting that 'to move swiftly, strike vigorously and secure all the fruits of victory, is the secret of successful war' (General Stonewall Jackson, cited in British Army, 2017: 8–15), and the other that destructing one's enemy's offensive capabilities contributes to victory (British Army, 2017: 8–19).

[21] Interestingly, for example, a division between levels of war that mirrors scholarly typologies of victory is deployed with reference to success, but not to victory.

they explain, 'is important because it provides a flexible context for actors to imbue concepts with their meaning, including both those who develop the concept as well as "audiences" who can internalize their meanings individually and even personally—thus, emotionally' (ibid.). When deployed to grasp empirical 'realities', however, the meaning of concepts gets fixed and at least part of their ambiguity is dispelled. In British doctrinal sources, however, it is arguably success that gets fixed, while victory is kept ambiguous—the well-known murky notion that scholars continuously try to grapple with. British armed forces should have 'war-winning capabilities' but, as per doctrinal indications, individual echelons are only responsible for—and potentially rewarded on the basis of—accomplishing success in their own building block. Even more interestingly, these narrower tasks lead to facts on the ground that a well-thought-out narrative—not necessarily articulated by the military, let alone those same echelons—might well change.

A purposefully murky notion of victory and a resolutely specific concept of success can be taken to work as a narrational strategy through which doctrine is made to concomitantly appeal to its varied audience. On this account, embedding the propositions about narratives *within* the nested model of success does not contradict the model, but consolidates it further, increasing its internal coherence. Imagined as victory's more manageable and less emotional sibling, success matches the rigid, hierarchical structures upon which the army is built. Enlistees are allotted narrow tasks and small building blocks. Commanders are similarly allocated operations—that is increasingly broader chunks of the overall picture—the overarching success of which is not fully within their province. The ultimate end state is political and at least partially dependent on a narrative that is not necessarily managed by the military. As a result, the possibility is put in place for the externalization of an overall failure, or defeat, that does not invalidate lower-level successes. After all, military operations set the stage for a success which 'is increasingly difficult to achieve in purely physical terms' (British Army, 2017: P1). And yet, armed forces are expected to have 'a will to win' (DCDC, 2017b: 29). But if winning is not entirely the same as succeeding, then victory can maintain its feature of an emotionally fraught notion that has allure across echelons. Working as an enticing promise, victory parallels army recruitment advertisements that suggest that 'failure' is 'the first step towards victory' because, by joining the armed forces, 'you fail, you learn, so you can win' (*Watch The Army's New Recruitment Adverts for 2021 | Forces TV*, 2021) or that the UK's is a 'winning army' whose members 'are paid to fight and win' (*Ready to Fight and Win | Who we Are | British Army*, 2019). And it is precisely upon such representation of victoriousness that the narrative hinges when it comes to addressing external audiences—be they aspiring recruits, or the wider public called to support them during war.

Reconsider the temporal work that the nested model of success carries out. While in the vernacular the notion of victory carries, as I said, connotations of finality, the idea of success does not necessarily do so. In fact, for Campbell and Steele (2021: 45) '"winning" and success are no doubt both ambiguous but also powerful signifiers in that there exists an assumed finality and clarity in the concept itself, thereby making it a socially *attractive* concept as well' (Campbell and Steele, 2021: 45). Beyond positing that the emotional resonance of victory *and* success is, at least partially, related to their association with finality, they also admit that 'both articulations of winning and success in war could be engaging slightly different concepts' and 'for sure ... engaging different *timings* or sequences within or before war' (Campbell and Steele, 2021: 47).[22]

The few references to victory which are found in historical vignettes throughout doctrinal texts reinforce the rupture-like temporality of the notion. In an illustration of the importance of achieving 'control of the air', JDP 0-30 *UK Air and Space Power* (DCDC, 2017c: 29) refers to the First Gulf War as a 'decisive victory with only a few coalition casualties'.[23] Similarly, JDP 0-10 *UK Maritime Power* traces the origin of the contemporary 'global rules-based international order' (one that 'the UK sits at the heart of' and that 'has provided stability to the world's political and economic landscape [*sic*]') to 'the end of World War II' (DCDC, 2017a: 11)—a claim which is reinforced in JDP 05 *Shaping a Stable World* by labelling the existing global order as characterized by a 'post-1945 international organisational structure' (DCDC, 2016b: 33).[24] Success, by contrast, cannot be equally time-altering—or more precisely, not all 'successes' within the model. Indeed, if lower building blocks are to feed into higher building blocks, in turn contributing to an overarching success, then only the last one—the 'end-state'—can have a rupture-like temporality. The temporality of lower-level objectives, that is, needs to be one of linear continuity. However, it is peremptory that success is *potentially*

[22] Working within the Just War Tradition, these authors suggest that 'success' is a prudential principle which is supposed to restrict war on the basis of the likelihood of its successful conclusion—that is, something to be assessed before embarking in the war effort. By contrast, 'winning', which they analyse within an academic literature on public support to warfighting, is a permissive principle which, still on the basis of victory's likelihood, enables the continuation of war. In this case, however, the assessment takes place once war already began and only to determine its protraction.

[23] Similar instances include the Duke of Marlborough's 'great victory at Blenheim in 1704 (British Army, 2017: 10–10) and Sir Edmund Allenby's 'stunning victory' at Megiddo, Palestine during the First World War (British Army, 2002: 3–3, 3–4).

[24] For an interpretation of 1945 as a less decisive victory, see the dispute around the Bronze Soldier of Tallinn in Chapter 5.

time-altering—i.e., that the sum of all building blocks provokes a temporal rupture, an end to wartime and a start to peacetime. Only in this way can the model be coherent, magnifying the affective economy sketched out above—i.e., audiences are allured by victory and managed through success. Also in this case, the differential use of victory/success concedes as much. Kept within the doctrinal apparatus via its past manifestation, victory keeps vivid the association with finality. Complementarily, finality is associated to success via the end state formulation above, and the acknowledgement that 'it is the triumph of the narrative that is decisive, not necessarily the facts on the ground' (British Army, 2017: i).

## Conclusion

Working closely with some of the doctrinal sources under analysis in this chapter, David Morgan-Owen and Alex Gould (2022: 19) have argued that doctrinal 'knowledge claims are inherently subjective, and that the mobilisation of expertise as a form of power within national security discourse and civil-military relations must therefore be viewed as a form of political activity'. Doctrine production, in other words, is a political process. Tracing an increasing trend within UK doctrinal sources to 'focus upon future war', Morgan-Owen and Gould (2022: 10), read recent doctrinal developments as agenda-setting exercises that hinge on the political capital associated with military expertise in the areas of strategy and defence.

In a similar vein, this chapter sought to illuminate the politics of British doctrine by focusing on the issue of military victory. It has done so by foregrounding the links between military developments and narratives that doctrinal sources articulate, and their relation to time. In particular, I found that UK military doctrine articulates a nested model of success in which victory is presented as the outcome of both facts on the ground and their narration. I demonstrated that a focus on time helps taking stock and making sense of this proposition. It also sheds light on its political and analytical implications. For example, this chapter has showed that through narrational slippages between 'victory' and 'success', British doctrine is able to 'mould temporality' and thereby concurrently appeal to the affective dispositions of different audiences.

The insights derived from this analysis of doctrine as a victory practice have important implications for the study of victory too. Firstly, to accept that victory is jointly constituted by war's material realities and their narration is to also accept that extant studies that focus on only one of these dimensions are, at best, partial. Rather than pertaining to separate domains concerned with different temporal phases, military developments and the political narratives articulated to make sense of them and communicate them to audiences are intimately connected. In fact, victory's material and narrational dimension interact and interlace, so much so that they can be conceived as co-constitutive.

Secondly, and relatedly, the study of British doctrine in this chapter suggests that victory research can be extended to complementary corpora of knowledge. If doctrinal thought contributes to the production, reification, or even contestation of victory's conceptual set-up, so do other metanarratives of victory, such as scholarly analyses of the notion or other, non-doctrinal, theories of warfare. How do they conceptualize victory? What insights can be gathered from these accounts in terms of victory's conceptual apparatus? Most importantly, how do the conceptual politics of victory play out in these alternative (meta)narratives? Answering these questions, as this chapter showed, already presupposes a double-layered analysis which is attentive to both the *content* of these conceptual insights as well as their *politics*.

Lastly, if this proposition is taken seriously, the analytical contours of social-scientific and historical inquiries on victory must be redefined. Consider, for instance, widely accepted victories of the recent past, like those of the First and Second World Wars. To be considered alongside weaponry, tactics, and military strategy in the analysis of these victories are the stories told—and that we keep telling—about those events. A study of these victories ought to concomitantly consider, for example, the politics of V-Day, from its inception—via, for instance, historical public speeches and announcements—until the present days, including the analysis of a huge memorialization industry committed to the continuous reproduction of those 'facts'. Most importantly, these victory practices—spanning from commemorative events and public celebrations, movies and TV shows, to history curricula in schooling—should not be approached as separate phenomena, or the upshot of historical 'facts' and 'events', but integral to them. It is precisely to examples of these practices that the next chapter turns.

# 3

# Textbooks, Poppies, and Videogames

## Victory-Making in the UK

## Introduction

> The cymbals crash
> And the dancers walk
> With long silk stockings
> And arms of chalk
> Butterfly skirts
> And white breasts bare
> And shadows of dead men
> Watching 'em there
>
> Shadows of dead men
> Stand by the wall
> watching the fun
> Of the Victory Ball
> They do not reproach
> Because they know
> If they're forgotten
> It's better so …
>
> 'Pish!' said a statesman
> Standing there
> 'I'm glad they can busy
> Their thoughts elsewhere
> We mustn't reproach 'em
> They're young you see'
> 'Ah' said the dead men
> 'So were we!'

**[Noyes, 1920]**

*Military Victory Beyond the Battlefield*. Mirko Palestrino, Oxford University Press. © Mirko Palestrino (2025).
DOI: 10.1093/9780198955818.003.0004

Reprinted in the *Daily Herald* on 11 November 1922 (Gregory, 1994: 88, fn. 45), Noyes' poem above epitomizes a widespread ambivalent attitude towards the First World War victory in the 1920s in Britain. In the poem, a cheering attitude of festivity is juxtaposed to the tragedy of death, highlighting the ambivalence of military victory: a supposedly joyful *event* worth celebrating that still took death and destruction to *achieve*. Dichotomous thinking surrounding military victory—and, in yet another dichotomy, defeat—is quite frequent. Foregrounding one of these victory-related binaries, the previous chapter dispelled a pervasive, problematic distinction between the materialities of warfare and their narration. It postulated that military victory is instead co-constituted by military and martial developments—or 'facts on the ground'—*and* various narratives articulated to make sense of them. Contextually, I also suggested that the research archive should be expanded as to encompass seemingly non-martial practices unfolding outside the battlefield that are similarly implicated in the construction of military victories.

This chapter turns to some of these and challenges yet another dichotomy associated with victory: wartime/peacetime. It focuses on the UK to simultaneously expand the empirical remit of this enquiry from the institutional to the national level. Homing in on what could be named the 'cultural production' of victory in the UK, I explore a variety of apparently unrelated practices through which victory is *routinely* produced: history teaching in (civilian) schools, war commemorations—especially Remembrance Sunday rites—and war-themed videogames. Contextual differences between them notwithstanding, these practices all consistently construe victory as a material event that takes place in a specific, demarcated space-time. The chapter argues that the construction of victory as an 'event' endowed with causal power lasts decades, if not centuries. In short, it takes ages for victory to happen in a day. While seemingly paradoxical, this tension is unsurprising. Just as any other timing outcome, victory appears to be self-evident, static, and circumscribed, but hides complex and enduring temporal operations (see Hom, 2018a, 2020).

I start by analysing GCSEs and AS/A-Level[1] textbooks currently employed in the UK for the teaching of history. I show that, also in civilian schooling, victory is made to coincide with the outcome of spatio-temporally limited battles or operations, and that even greater emphasis is put on the causal power of these military developments. Among the political implications of these dynamics, however, is a simplistic chronological rendering of history implicated in the effacement of Britain's colonial pasts. The chapter then

[1] The General Certificate of Secondary Education and the Advanced Subsidiary/Advanced level qualifications are the most common and widely recognized pre-university qualifications in the UK.

turns to revisiting the history of Remembrance Sunday.[2] I show that facts on the ground were far from being decisive in 1918 and 1919, and that victory celebrations were instrumental to the engineering, or creation, of victory—made to coincide, in Britain, with 11:00 a.m., on 11 November 1918. The consolidation of these extremely specific temporal coordinates still hinges, to date, on the routinary performance of these over one-hundred-year-old rites. I conclude the chapter with the analysis of popular war-themed videogames. Both First Person Shooter (FPS) games like *Medal of Honor* and *Call of Duty* and narrative-focused adventure games like *11:11 Memories Retold*, partake in the production of military victories that are temporally bounded and essentially material in character. This idea is embedded in videogames either by design—i.e., constructing games that advance from one level to the other upon completion of fighting- or killing-related tactical and operational objectives—or by reinscribing character narratives within a broader conflict that is presented as spatio-temporally demarcated.

## Victory's causes and victory as cause: the teaching of victory in British history education

'In a way that seems true of no other subject' David Cannadine (2013: 14) writes, 'the issue of history teaching in schools engages public attention and carries unmistakable political overtones: for unlike, say, economics or physics, where the content is similar the world over, classroom history is in large part about the national past, and thus about who we are, where we have come from, and what sort of country we are living in'. This and other similar commentaries filled the pages of newspapers and other British media between 2011 and 2014 in reaction to a proposed review, and ensuing reform, of the history national curriculum—sponsored by the then Tory Secretary of State for Education Michael Gove (see Alexander and Weekes-Bernard, 2017: 478–484). Observers and academics alike denounced the overtly Anglocentric character of the new history curriculum (Cannadine, 2013), its narrow focus on 'Fundamental British Values' (Lidher, McIntosh,

[2] Armistice Day and its successor, Remembrance Sunday, are arguably one of the most widely studied commemoration practices to date. Scholarship on the subject has been growing steadily throughout the past century, with a sharp increase in research registered around the centenary of the First World War (Andrews, 2015: 104). Historians and social scientists alike have explored First World War–associated rituals, rites, and celebrations far and wide, providing insights into the gendered, racialized, and militaristic dimensions of these practices (Basham, 2016; see also Mycock, 2014); their links with religion (Gregory, 1994; Hukelova and Holloway, 2021), schooling and education (Haight et al., 2021), and even sports (Kelly, 2023). This abundance of scholarship notwithstanding, the link between victory, temporality, and Remembrance Sunday has been largely overlooked. However, as this chapter demonstrates, this association is particularly salient.

and Alexander, 2021: 4223), as well as 'the potentially homogenizing vision of "Our Island Story"' (Bowen et al., 2012: 126). The new curriculum, in other words, was often found to be excessively nationalistic in character, and complicit—among other things—to 'Britain's historical amnesia' of some of its darkest pages of colonial history (Lidher, McIntosh, and Alexander, 2021: 4224–4229), as well as to the erasure of 'migrants, black and minority ethnic and religious communities' (Alexander and Weekes-Bernard, 2017: 482; see also Wilkinson, 2014) from history textbooks.

Public and academic debates of this kind are nothing new in the UK. The introduction of the first, government-approved National Curriculum in 1988 (Dunn, 2000), for instance, exacerbated controversies over the content and politics of mandated history programmes that were felt to reflect ideological commitments to the New Right (see Crawford, 1995). In fact, history teaching is (in)famously the object of political disputes worldwide. Heated discussions known as 'history wars' (see Taylor and Guyver, 2012) are increasingly frequent across the globe, and scholarly consensus over the implication of schools in the construction and popularization of canonical national narratives of the past is widespread (see, for example, Lidher, McIntosh, and Alexander, 2021: 4222). For instance, while leading to very different outcomes, history education played a central role in the (re)construction of national identity narratives in both post–Second World War Japan and Germany (Dierkes, 2010). Similarly, the Vietnam War powerfully shaped history teaching in the US (Loewen, 2000; see also Griffen, 1974). In these countries, military defeat worked as a key organizing principle for the creation of a history curriculum that could help grappling—in one way or another—with difficult national pasts (Dierkes, 2010; see also Hein and Selden, 2000).

This relation between military defeat (and victory) and history teaching, however, is not unidirectional. Indeed, not only is the moulding of history curricula a convenient strategy to design and promote narratives that help polities cope with military victories and defeats, but it also helps constructing those victories and defeats in the first place. Paralleling the study of doctrine in Chapter 2, this section starts enquiring into the making of victory in British culture by treating history teaching in school as another knowledge production process that—I argue—is central to the construction of victory and its normalization as the material outcome of armed conflicts. In particular, I survey a variety of GCSEs and A-Level history textbooks currently employed in the UK.[3] Not unlike their military counterparts, pupils in civilian schools are

[3] At the time of writing, GCSE, AS, and A-Level qualifications in the UK are awarded by four, institutionally approved exam boards: AQA, OCR, Pearson, and WJEC Eduqas. While the Department for Education has responsibility over curriculum and subject content development, the exam boards are ultimately in charge of developing and awarding the qualifications (see https://www.gov.uk/guidance/

taught that victory is a time-altering event that takes place *during* and *through* war and brings conflicts to closure. History education in the UK, in other words, reiterates the idea that victory correlates with finality—popularizing the falsely dichotomous distinction between war's materiality and narratives (and/or wartime and peacetime) problematized in the previous chapter. In fact, history textbooks are even more stark in presenting victory as one of the material outcomes of conflict, and one which has significant causal power—i.e., it engenders sociopolitical and historical change. Taken together, these moves help reiterate the idea that victory is solely the (material) outcome of armed conflicts and obscure victory's narrational dimension.

When understood within the context of history teaching as a fundamentally political practice, this peculiar narration of victory acquires crucial importance. As UK debates on the politics of the National Curriculum vividly suggest, history teaching is also history-making. On the one hand, by equating military victory to the material outcome of conflicts and ascribing causality to martial events alone, these textbooks contribute to the proliferation of a well-known, almost ahistorical understanding of war as an exceptional, temporary business. This idea, however, has been repeatedly underscored to be permissive of violence (see Dudziak, 2012), unfaithful to lived experiences of war (see Chapter 4), and complicit to the erasure of—among others—colonial violence (Barkawi, 2016) and gender-based violence (Sjoberg, 2013). On the other hand, by depicting victory as the sole product of military developments, attention is detracted from the violence produced by and through political narratives.

## Victory and causation

In the introduction to *Power and the People*—an Assessment and Qualifications Alliance (AQA)-approved GCSEs history textbook[4]—Alf Wilkinson (2016) writes:

regulating-gcses-as-and-a-levels-guide-for-schools-and-colleges-2022/about-gcses-as-and-a-levels). As a result, exam boards provide detailed information about the approved educational material—and especially textbooks—to be used in preparation of the relevant qualification. The analysis in this chapter is based on AQA's approved GCSEs and AS/A-levels history textbooks, available at https://www.aqa.org.uk/subjects/history/gcse/history-8145/specification/specification-at-a-glance and https://www.aqa.org.uk/subjects/history/as-level/history-7041/specification/specification-at-a-glance respectively.

[4] At the time of writing, AQA is the most widely employed exam board in the UK for both GCSEs and AS/A-Levels—setting 'papers for more than half of all GCSEs and A-levels' (AQA, n.d.). I chose to focus on textbooks from a single exam board for three reasons: (a) coherence in the corpus of texts under analysis, (b) convenience in terms of accessing these sources (most AQA-approved textbooks are available for consultation at the British Library), and (c) AQA's unparalleled popularity when compared to the other exam boards. I do not wish to claim that the analysis is representative of all GCSEs and AS/A-Level history textbooks currently employed in the UK, nor to have produced 'generalizable' findings, representative of a monolithic 'British historical pedagogy'. Instead, the sources are employed to provide meaningful

> This book looks at the history of Britain from around 1200 to today largely through the eyes of people who thought something was wrong and wanted to put it right ... This is **the story of power**—of who runs the country, about who makes the laws, who is in control. This is also **the story of protest** ... In one sense, it is also **the story of the spread of democracy**.
>
> (Wilkinson, 2016: Introduction; bold text in the original)

The book is one of three 'thematic studies' textbooks that students must choose to complete their GCSE history qualification,[5] and it sets itself up to explore over 800 years of British history through the lens of 'power'. Students are guided through some key historical events and figures over four main historical periods—'the middle ages', 'the early modern world', 'the nineteenth century', and 'the twentieth century'—and invited to focus, throughout, on the differences between power distribution among individuals (like the king or the pope) and groups (from barons, to parliament, women, and workers)—asking questions about who holds the most power and when, and what led to these changes. Perhaps unsurprisingly, given the focus on people and power, students are introduced to military victory (and defeat) early on in the textbook and several times throughout. Most interesting, however, is the consistent, self-evident, and unreflective deployment of victory to refer to the *material* outcome of conflict—often reinforced by vivid descriptions of battles and military tactics—as well as the ascription of temporal finality and even causality to the notion.

After a brief context note on 'Medieval Britain', the textbook dives right into the history of the Magna Carta, one of the key developments in British medieval history when it comes to the distribution of (political) power (Wilkinson, 2016: 1). 'War and money—and military *defeat*', Wilkinson (2016: 1.2) suggests, 'are at the heart of our story'. According to the textbook, indeed, it was a military defeat that led King John to sign the Charter:

snapshots, or vignettes, of educational material currently employed for the teaching of history in support of my theoretical argument.

[5] The subject content for GCSE History awarded by AQA is divided into two 'papers', each of which comprises two sections. The first paper looks at international history, and comprises a 'period study' section—America 1840–1895, Germany 1890–1945, Russia, 1894–1945, and America 1920–1973—and a 'wider world depth study' section—to be chosen among (a) The First World War, 1894–1918; (b) The inter-war years, 1918–1939; (c) Conflict and tension between East and West, 1945–1972; (d) Conflict and tension in Asia, 1950–1975; and (e) Conflict and tension in the Gulf and Afghanistan, 1990–2009. Students must choose one text from each section. The second paper, instead, has a national focus, with a first section on 'thematic studies', and a second section on 'British depth studies'. Three options are available for the former, namely: (a) Britain: Health and the people: c1000 to the present day; (b) Britain: Power and the people: c1170 to the present day; and (c) Britain: Migration, empires and the people: c790 to the present day. 'British depth studies', instead, comprises four options: (a) Norman England, c1066–c1100; (b) Medieval England: the reign of Edward I, 1272–1307; (c) Elizabethan England, c1568–1602: and (d) Restoration England, 1660–1685. Also in this case, students pick one text for each section among the available alternatives. For more information visit: https://www.aqa.org.uk/subjects/history/gcse/history-8145/specification-at-a-glance.

> In June 1215 two armed camps faced each other across the meadow at Runnymede, next to the River Thames. One was made up of most of the leading barons in the country; the other, King John and his few remaining supporters. Months of tension had led to this moment. Now, faced by a superior army King John felt he had no choice. He must sign the Charter.
>
> **(ibid.)**

The reasons for these tensions are, once again, intimately linked to a military defeat. For Wilkinson (ibid.), the victory of the French king over Normandy in 1204 had led King John to 'spen[d] the rest of his rule trying to win it back', and to use 'every means at his disposal to collect money to pay mercenaries to fight for him in France'. Among these was the repeated increase in taxes for knights and barons—a choice that made the sovereign profoundly unpopular. However, '[t]he final straw for the barons', the textbook explains, 'was another defeat in France in 1214, at the Battle of Bouvines' (ibid.). This new defeat exposed King John's ultimate incapacity to win Normandy back and drove the barons to a tipping point: facing the sovereign militarily and force him to sign a charter. Two familiar victory tropes emerge starkly from this account: military victories (or defeats) are (a) crucial *events*—i.e., temporally bracketed, material, and martial phenomena, seemingly taking place in a circumscribed time-space (a wartime battlefield)—that (b) leave fundamental marks on historical chronologies—including those of not immediately martial phenomena, like the Magna Carta.

Indeed, not only does the textbook recall the history of the Magna Carta as a concatenation of military victories and defeats, but it also encourages students to identify causal relationships between events. King John lost Normandy to the French King, in 1204, a defeat—i.e., a circumscribed, material, and clearly identifiable event—that led him to use every means at his disposal to win that territory back. This military defeat, in turn, lies at the heart of John's choice to raise increasingly high taxes, *causing* the barons' discontent with the sovereign's rule. Another military defeat, in 1214, *engenders* the barons' uprising *and* the signing of the Magna Carta. In short, victories or defeats are key *causal factors* moving history forward.[6] A 'focus task'[7] on the same topic, for instance, asks the students to identify '[w]hat factors led to the revolt of the barons and the signing of Magna Carta', pre-identifying war as a

[6] Importantly, victories and defeats are not the *only* factors that the textbook identifies—nor do I wish to claim they are always and necessarily the most important ones. In relating the history of the Magna Carta, for example, Wilkinson (2016: 1.2) suggests religious development also played a role. My claim, instead, regards the consistent deployment of victory (and defeat) as a (material and spatio-temporally limited) causal factor.

[7] A specific exercise 'that appears at the end of each topic or at the end of each chapter' designed to get students to think about what they have learnt and build their personalized 'big picture of the story of power and protest' (Wilkinson, 2016: How this book will help you). This is a standardized exercise that can be found across other AQA-approved GCSE history textbooks in the same series.

key, decisive factor and highlighting the importance of 'John's failures in war' (Wilkinson, 2016: 1.2). Similarly, a 'progress check' exercise requires learners to identify two 'causes' that led to the institution of the Magna Carta, prompting them to 'explain why the barons were unhappy with King John' and 'why the barons revolted in 1215 and not earlier'—both questions intuitively bringing the military defeats of 1204 and 1214 to the forefront.

The following topic that the textbook explores, '[t]he origins of Parliament', reiterates this logic. Wilkinson (2016: 1.3) opens the discussion of another, seemingly non-martial, key historical development with a vivid description of a military defeat:

> It is 8.30 a.m. on 4 August 1265 and two armies, one much bigger than the other, gaze at each other just a mile outside of Evesham in Worcestershire. One is led by Prince Edward, son of Henry III, and is made up of an estimated 18,000 knights and infantry. The other is led by Simon de Montfort, ruler of England for the last twelve months, who is holding the king captive. Simon is vastly outnumbered but leads his knights in a desperate charge to split up and destroy the royal cavalry. It succeeds initially, but Simon and his knights are quickly surrounded and mercilessly killed … This defeat effectively ends Simon's rebellion against the King, even though it takes another two years before Henry III's kingdom is totally at peace.

The chapter proceeds by recalling the history of parliament backwards, suggesting that two major French victories in 1230 and 1242 had led Henry III to rule similarly to his father John, asking for increases in taxes by summoning 'King's Councils'—or 'Parliaments' (ibid.). After siding with 'the knights and commoners' in one such inconclusive meeting, Simon de Montfort brought the issue to the battlefield—a story that culminated with 'Simon's victory at the Battle of Lewes in May 1264' and that led to a year-long rule terminating in the Battle of Evesham that opens the chapter. 'Nevertheless', Wilkinson (ibid.) concludes, 'a precedent had been set. Henry III and later Edward each in turn felt the need to summon the Commons to Parliament and get their consent to taxation. Perhaps that is why Simon de Montfort is regarded as the "Father of Parliament"'. Once again, then, victories and defeats appear as discrete events fuelling the engine of history: they are material causal forces bringing about change.

Such an understanding of victory is not topic- nor textbook-specific. Rather, victory is consistently presented as a material event endowed with causal power across historical periods, topics, and textbooks. When analysing 'the English Revolution' in the same textbook, for example, Wilkinson (2016: 2.3) describes the Civil War in 1642 as a central event and stand-off in which 'neither side—King nor parliament—could summon enough resources to defeat the other until, in 1645, the New Model Army ... made the decisive breakthrough'. In other words, a predominantly material martial development (a new army model) leads to a spectacular victory[8] that, in turn, engenders the King's execution, the abolition of monarchy, and the establishment of a commonwealth.[9] Analogously, Wilkinson (2016: 4.1) suggests that '[p]erhaps it is war that has moulded society more than anything else throughout the twentieth century' and that 'conscription'—a phenomenon that had 'sent men off to fight and women to work in [new] occupations'—'played a vital part in securing victory for Britain in both world wars'. Across cases, then, the timing indexicals 'victory' and 'defeat' conceal operations of ordering in which tactics, battles, and other military developments are the key factors being related. While this selective narrative operates as the timing standard, both victory and defeat emerge as timing outcomes that are predominantly—if not wholly—material in nature. In turn, these 'events' are embedded in broader narratives as key determinants of historical change and continuities.

Notably, the same logic is reproduced in other textbooks. Explaining the expansion of the US in the nineteenth century, Ferriby, Martin, and Walsh (2016: 1.1) adopt an analogous notion of victory, suggesting that victory in 'the Battle of Buena Vista'—in the Mexican–American War, 1846–1848—changed the US attitude towards 'the Great American Desert' and ultimately led to expansions and settlements in the Indian lands known as the Great Plains. On a similar note, the textbook identifies the Prussian victory against France in 1870 as the origin of the unification of Germany (or the German Empire) in 1871—Ferriby, Martin, and Walsh (2016: 2.1, my emphasis; see also 5.1) write: 'Germany had only been a united country since 1871. Before that what we call Germany had been made up of several independent states, the most important of which was Prussia. This new empire was created in 1871 *after victory* against France in 1870'. Again, and in another different textbook, the narration of the Vietnam War also revolves around key military victories and defeats such as the military occupation of French Indochina by Japan

[8] Wilkinson (2016: 2.3) writes: '[t]he army had an immediate military impact and it was the decisive factor in winning the war for Parliament'.

[9] Unsurprisingly, the 'focus task' for the topic also identifies 'war' as the most important factor 'in the English Revolution' (Wilkinson, 2016: 2.3).

and the Vietminh's victory over Japan (Bruce and Wilkes, 2017: 32–33); the Tet Offensive in 1968—explicitly identified as a turning point (Bruce and Wilkes, 2017: 60, 82) leading to 'American defeat' (Bruce and Wilkes, 2017: 86); and the fall of Saigon in 1975. The narrative is corroborated by long explorations of both sides' military tactics (Bruce and Wilkes, 2017: 52–61).

This coherent treatment of military victory (and defeat) as a circumscribed, material event and its explicit association with causality across textbooks and historical periods is perhaps unsurprising. The intended learning outcomes and subject content for GCSEs—set centrally by the Department of Education (see Department for Education, 2014c: 3–12)—expect students to 'develop and demonstrate knowledge and understanding of key features and characteristics of the periods studied: the *chronology*, individuals, *events*, developments and issues in the specified content' as well as show an 'understanding of key features and characteristics of the periods studied in relation to second order historical concepts of *continuity*, *change*, *cause*, *consequence*, significance and similarity and difference within situations' (Department for Education, 2014b: 6; my emphasis). In other words, the institutional guidance for the teaching of history at GCSE level emphasizes the importance of identifying historical events, learning and establishing chronologies, and detecting causes and consequences of historical continuities and changes—a focus that AS and A-Level specifications express explicitly in terms of demonstrating an understanding of 'causation' (see Department for Education, 2014a: 2).[10] Moreover, and as I stress throughout this book, in the vernacular, victory tends to be associated with the military realm in general and the end of wars in particular.

Arguably, then, it is both pedagogically and logically sound that textbooks single out specific victories and defeats as relevant historical events and encourage pupils to make connections between these and wider sociopolitical change. The issue with this approach to victory, however, is that it risks collapsing the narrational and material dimensions of victory discussed and problematized in the previous chapter. The problem, here, is twofold. Firstly, like contemporary doctrinal texts in the UK, these textbooks also obscure the relationship of mutual constitution that links victory's material and narrational components by attributing primacy to the former over the latter. While narratives play a crucial role in the timing operations underpinning

[10] For AS and A-Level learning outcomes and subject content see Department for Education (2014a). Both GCSEs and A-Level specifications for history explicitly build on Key Stage 3 learning outcome and subject content (Department for Education, 2014c: 94–97). Both the focus on chronology and events, change and continuity, as well as cause and consequence, is consistent throughout.

victory and defeat, they remain hidden. The impression they give, in other words, is that victory is the result of military choices and tactics *alone*—some sort of self-contained, technical achievement. Secondly, and because of this collapse, the idea of causality attributed to the notion is metonymically passed onto victory's material component, so that it is only the facts on the ground that determine victory and, through it, other sociopolitical consequences—put differently, martial developments work as a victory symbol.

The political implications, here, are salient. For example, this understanding of victory risks contributing to an unsophisticated chronological rendering of history that only focuses on 'kings and queens, ministries and battles, *war and peace*', is 'decidedly blinkered in its scope', and can only be taught in 'patchy, simplistic, superficial and disconnected ways' that 'abandon any pretence that history is about understanding as well as about knowing' (Cannadine, 2013: 15; my emphasis). Unsurprisingly, the chronological unfolding of history as a relentless, causal chain of events seemingly at work here has been found, for example, to feed directly into the 'Our Island Story' and the 'Fundamental British Values' agendas (Lidher, McIntosh, and Alexander, 2021). Indeed, linear chronologies of events tend to 'downplay the internal diverse histories of "our Island story", placing them as "out there" rather than "in here" and of relevance "then" rather than "now"' (Lidher, McIntosh, and Alexander, 2021: 483). Moreover, oversimplification of complex historical dynamics remains a risk.

An obvious case in point is the treatment of victory in the 'Great War' across relevant GCSE and A-Level textbooks.[11] Ferriby, Martin, and Walsh (2016), for instance, devote an entire chapter—titled 'Conflict and Tension, 1894–1918'—to the First World War, exploring the topic in depth. The chapter, they write, 'focuses on the causes, nature and conclusion of the First World War. The war brought huge changes to society, to politics and to the economies of leading world powers' (Ferriby, Martin, and Walsh, 2016: 5). A visual timeline recalls the conflict's key events between 1914 and 1918 as follows: (a) 1914 Assassination of Archduke Franz Ferdinand leading to the First World War; (b) 1915 Gallipoli Campaign; (c) 1916 Battles of Verdun and

[11] Among AQA-approved history textbooks for both GCSEs and AS and A-Levels, there are four main books covering the First World War in depth, namely *Understanding the Modern World* (Ferriby, Martin, and Walsh, 2016) and *Conflict and Tension: First World War 1894–1918* (Wilkes, 2018) for GCSEs; and *War and Welfare: Britain in Transition 1906–1957* (Willis and Thomas, 2016) and *International Relations and Global Conflict c1890–1941* (Kearey, 2015) for A-Levels. Accordingly, this discussion focuses mainly on these four texts, while also making sporadic references to other textbooks that treat the 'Great War' tangentially—e.g., as one phenomenon within a longer historical trajectory.

the Somme; Battle of Jutland; (d) 1917 USA enters the war; Battle of Passchendaele; (e) 1918 November—Germany surrenders; armistice (ibid.). An equally—if not more—military-focused narrative characterizes the chapter.[12]

The textbook starts off by providing a description of 'the alliance system' with brief blurbs of key actors on each side,[13] a quick focus on the 'Anglo-German rivalry', their 'naval race', and the issue of 'European rearmament'; brief explanations of 'pre-war crises'—namely 'Morocco, 1905 and 1911' and the 'Balkans'—follow; it then identifies the 'assassination of Archduke Franz Ferdinand' as the *casus belli* and turns to the 'Schlieffen Plan' (Ferriby, Martin, and Walsh, 2016: 5.1). The central section of the chapter tackles the conflict itself, and is overwhelmingly focused on martial conditions and events, military tactics, and weaponry (Ferriby, Martin, and Walsh, 2016: 5.2). Here, a recollection of some key battles (the same in the timeline above) and the military tactics employed in each of them (e.g., naval warfare, trench warfare, use of artillery, etc.) is followed by discussions of soldiering (life in the trenches) and weaponry (artillery, machine guns, poison gas, tanks, and aircraft); the section is concluded by another recollection of battles (ibid.).[14] The concluding section of the chapter finally considers the end of the war, dealing 'with the reasons for the Allied victory in 1918 following the German surrender' (Ferriby, Martin, and Walsh, 2016: 5.3). That the material component of victory is privileged and attributed causal power is already evident from the section's outline:

> In this part of the topic you will study the following:
>
> - Changes in the Allied forces, why the USA entered the war, and the impact of these developments on the Allies' chances of victory.
> - Military developments in 1918 and their contribution to Germany's defeat.
> - The German surrender and why Germany asked for an armistice.
>
> **(ibid.)**

Unsurprisingly, the end of the war is explained in terms of a concatenation of military developments 'on the ground':

[12] The other Great War focused GCSE history textbook, *Conflict and Tension: First World War 1894–1918* (Wilkes, 2018) adopts an analogous structure.

[13] Namely Germany, Austria-Hungary, and Italy for the Triple Alliance (or the Central Powers), and Britain, France, and Russia for the 'The Triple Entente' (Ferriby, Martin, and Walsh, 2016: 5.1).

[14] Importantly, the categorization of most of these battles as victories, defeats, or stalemates is provided. This is mainly based on factors such as weaponry, tactics, and technology (see Ferriby, Martin, and Walsh, 2016: 5.2).

> Between May and August [1918] the Germans made no further progress ... By now, [the Allies] had large numbers of well-fed and well-equipped troops, much of the equipment and manpower coming from the USA. These troops were supported by tanks, aircraft and improved artillery. By 1918 the big guns were capable of hitting targets with impressive accuracy as well as laying down smokescreens or giving covering fire for attackers. On 8 August the Allies counter-attacked along much of the Western Front. This later became known as the German Army's 'Black Day'. It was now just a matter of time before the Allies defeated Germany. By late September they had reached the Hindenburg Line. By October the Germans were in full retreat ... By November 1918 Germany's allies were surrendering.
>
> **(ibid.)**

The authors proceed by relating the 'abdication of the Kaiser', the establishment of the Republic, and the signing of the armistice:

> [t]he armistice was signed at 5 a.m. on 11 November, to come into effect at 11 a.m. Paris time (12 noon German time). Those signing the armistice were later accused of stabbing Germany in the back. They had no choice, but the army generals did not necessarily view it in this way. Across Europe, in both victorious and defeated nations, there was relief that the fighting was over. In Britain and France there was naturally much rejoicing. Discussions started in January 1919 at Paris about a detailed peace settlement, leading to the signing of the Treaty of Versailles in June 1919.
>
> **(ibid.)**

Three aspects of this military-focused account of the Great War are particularly salient. Firstly, victory is presented as a logical and linear consequence of exclusively martial developments: superior weaponry and tactics led to military advantages and ultimately to victory. Victory, in other words, is a technical business, the logical outcome of a well-fought war. Notably, the narrative advanced by A-Level textbooks is (expectedly) more complex and considers additional factors such as individual states' forms of government (Kearey, 2015: 107–111; Willis and Thomas, 2016: 43–51), the power of specific ideas—like self-determination (see Kearey, 2015: 107–111)—and personalities—like Lloyd George (see Willis and Thomas, 2016: 45–53). Nevertheless, a certain privileging of war's materialities is still apparent.[15]

[15] Kearey (2015: 110), for instance, also explains the end of the war as a concatenation of military developments led by the US intervention and the unsuccessfulness of German campaigns in the summer of 1918. In other words, even when other non-immediately martial factors are explored, they are brought back into a narrative that prioritizes 'the material' side of warfare—and considers, for instance,

Secondly, victory is once again a spatio-temporally circumscribed event—just like other victories of the past explored above. It clearly happened on 11 November 1918, at 11:00 a.m. Paris time. Again, this also follows on logically from the assumption that victory is the result of a series of military developments. Moreover, the point is consistently reiterated across textbooks, especially via visual timelines and graphic chronologies of different types, all marking 1918 (or 11 November 1918) as 'the end of the First World War' (see, for instance, Kearey, 2015: x–xi; Longley, 2016: 6–7; Willis and Thomas, 2016: x–xi, 44). Lastly, victory—and especially 'victory on the ground'—is once again treated as a causal force. On account of their victoriousness, the Allies set out specific terms for an Armistice that also led, among other things, to the dissolution of two empires and the establishment of the League of Nations (see, for instance, Kearey, 2015: 121–126).

Such an approach to victory in general, and victory in the First World War in particular, is at best simplistic and at worst partial or even incorrect. As I show in the next section, victory in the 'Great War' was far from being apparent in November and December 1918 and for a good part of 1919. In fact, a certain need to 'make' that victory—in the sense of consolidating, in the mind of the British people, the idea that the First World War was over and had been won—was felt throughout 1919 and the early 1920s. This led to the creation, consolidation, and normalization of a set of commemorative and celebratory practices that could instantiate and support such a victory narrative—a narrative that still hinges on its continued reproduction to this day. In other words, the notion of a clear-cut and decisive ending to the First World War at 11:00 a.m. on 11 November 1918 is mythical. That victory—as any other—is co-constituted by narrations of victory as a material and spatio-temporally bounded event of the kind that GCSEs and A-Level history textbooks reproduce. Complementing the study of doctrine in the previous chapter, the analysis in this section has sought to identify history teaching as another, and apparently non-martial realm, in which victory gets constituted, outside the battlefield and during peacetime. I now turn to yet another realm in which analogous dynamics are at play: commemoration practices.

the impact these factors had on the fighting. In discussing the armistice, for instance, she writes: '[a]rmies will generally *only* surrender if they have no other option and *cannot fight any longer*' (Kearey, 2015: 111).

## From Armistice Day to Remembrance Sunday: the making of victory in the 'World War(s)'

On the eve of 14 January 2013, with planning for the First World War centenary commemorations in Britain well underway, an editorial by Robert Hardman (2013) was published in the *Daily Mail* with the provocative title 'Why we SHOULD upset the Germans—by reminding them of their Great War atrocities'. For Hardman, commemorations of the Great War had been excessively focused on collective mourning, remembrance, and the solemn condemnation of war, to the detriment of celebrating what Britain has 'actually achieved' through that war: stopping German atrocities. The crux of the problem is that the UK has failed at commemorating victory. He writes:

> Oh dear. I think we can already see how this one is going to play out: mud, futility of war, lions led by donkeys, a bit of poetry, a nod to the nurses and the munitions girls and a solemn conclusion that it must 'never be allowed to happen again' ... As for history? Well that seems to be secondary. *But let no one mention the V word.* Indeed, you can scan the official blurb about the centenary and *the word 'victory' is barely mentioned.* No one 'won' this war, is the subliminal message. It just stopped because everyone had had enough.
>
> **(Hardman, 2013; my emphasis)**

The *Daily Mail* is not the only newspaper to articulate this concern and to do so by framing it precisely as a 'victory problem'. A *Sunday Times* article from the day before—titled 'Don't mention that we won the First World War' (Hellen and Brooks, 2013)—expresses similar concerns:

> 'There is an intent in government not to upset the Germans,' said a senior source who is involved in the centenary commemorations of the 1914–18 conflict. The high-level committee planning the ceremonies ... is split on the question. Sebastian Faulks ... is resisting pressure to celebrate a military victory in the way the Second World War is remembered. 'This must not be VE Day-like,' Faulks said ... One member, Professor Sir Hew Strachan, an eminent Oxford historian, argues that Britain should celebrate victory and also commemorate lost lives.

The piece proceeds by quoting Hew Strachan[16] at length on the point that 1918 'Germany was a militarist and imperialist regime which had to be defeated and from that point of view the victory is a serious victory', adding

[16] Hew Strachan left the University of Oxford in 2015 and is currently Professor of IR at the University of St Andrews.

that 'the draft official programme marks mainly the battles that have become a byword for senseless sacrifice and military stalemate' (ibid.). A comment on the importance of including the Battle of Amiens to the programme is also included: 'It ended the stalemate on the western front and, after the 100 days' offensive by the allies, led to the armistice'.

Three key themes emerge from these articles. First, the language of victory is oppositional. To explicitly talk about victory, or to rejoice in its celebration, is to antagonize 'the Germans' and to remind them of the war atrocities committed during the conflict. Second, victory is both the direct consequence and cause of material developments—what doctrinarians would refer to as 'facts on the ground' (see Chapter 2). It is the Battle of Amiens that tipped the scales and led to victory. And it is victory itself, in the form of the Armistice of Compiègne, that stopped the Germans and their atrocities. Third, victory is a clearly identifiable event, happening on a specific date and, one might add, a specific time: 11 November 1918, at 11:00 a.m.

Notably, these themes are not new—nor is the framing of the issues concerning the First World War commemorations as a 'victory problem', or the existence of a 'Victory in/and the Great War' debate. In fact, a few iterations of similar public discussions are identifiable throughout the past century. The question of commemorating 11 November as a victory rather than a special occasion for collective mourning, for instance, lies at the heart of the creation of commemoration rites in 1919 and the early 1920s—the current, normalized hegemonic framework of Remembrance Sunday as a mourning occasion notwithstanding. Taken together, I suggest, the themes raised in this recent public debate are indicative of the way in which victory (in the Great War) was, and keeps being, constructed. The contemporary emphasis on remembrance and mourning notwithstanding, Remembrance Sunday is historically tightly related to the idea of victory. In fact, Armistice Day—that is, the original name of Remembrance Sunday—was, and is some way still is, a victory celebration, and a very successful means for the creation, constitution, and continuous reproduction of victory in the First World War.

## Engineering victory: a British history of Armistice Day 1919

The first official celebration of 'Armistice Day' took place on 11 November 1919 and centred around a key, fortuitously created ritual that has persisted steadily until the present day: the two-minute silence (Gregory, 1994: 8–12; see also Homberger, 1976: 1430). To understand this practice—as well as other associated rites and Armistice Day celebrations more broadly conceived—it is vital to briefly consider the historical sociopolitical context within which it emerged.

Since the ceasefire in the autumn of 1918, Britain had been caught up in a general climate of incertitude and the government was actively trying to work on 'the public mood' at home (Homberger, 1976: 1429). Indeed, it was not clear that the war had necessarily ended on 11 November 1918, for at least two reasons. First, the fighting had not ceased everywhere in Europe, and conflicts were still ongoing, for example, in Romania, Hungary, Poland, Czechoslovakia, and Russia—or as Churchill put it, 'the wars of the pygmies have now started, the wars of the giants have ended' (Churchill, cited in Cooper and MacMillan, 2009: 124). Second, the 'giants' war' was not technically over either, or not until the Treaty of Versailles was officially signed on 28 June, 1919. As a result, a general fear that the conflict could re-escalate to full-on war again was widespread among various members of the British political elite. Historian Eric Homberger (1976: 1429) writes: 'Wilson warned Lloyd George and Sir Eric Geddes in January 1919 that the war was far from being over and that they "must prepare the public mind for armies of occupation in India, Gibraltar, Malta, France, etc". Churchill was pushing for intervention against the Bolsheviks in Russia'.[17] He adds that the fast demobilization of the army was also proving to be a major issue, and that the government was determined to 'win the hearts and minds of the public' (ibid.).

Possibly, then, the Cabinet had this objective of galvanizing the British people's spirit in mind when they appointed Lord Curzon as chairman of a Peace Celebrations Committee to plan for national festivities in celebration of the victory. Allan Greenberg (1989: 6), for instance, writes that 'the British government's proposal to sponsor a national day of festivities to celebrate the signing of the peace treaty then being negotiated ... was intended to divert public attentions from serious national problems that included an economic recession, labor unrest, and, even more threatening, manifestations of discontent in the army'. The time frame of the Committee's appointment, as well as their working schedule for the celebrations' planning, corroborate this narrative. Convocated in the spring, the Committee held its first meeting on 9 May 1919, 'with no idea when the Germans would agree to the terms on offer at Versailles' (Homberger, 1976: 1429; see also Greenberg, 1989: 6). As a result, the Committee 'had to plan for some unspecified date later

[17] Importantly, such an assessment is resonant with the work of other historians. Commenting on the general climate in 1918–1919, Margaret MacMillan suggests, for instance, that '[i]t wasn't like the Second World War, where it was quite clear who the defeated were and quite clear victors and the thing was over ... In 1918 and 1919 by contrast there was a sense that it wasn't all over and there was a sense that things had not settled down' (Cooper and MacMillan, 2009: 126). On the end of the First World War see Gerwarth (2016).

in the summer' (Homberger, 1976: 1429), tentatively settling on a four-day program of festivities to be held on 2–5 August (Greenberg, 1989: 6). The provisional programme included plans for various religious services, popular and civic festivities, a victory march, and a river pageant (Homberger, 1976: 1429; Greenberg, 1989: 6). The signing of the Treaty of Versailles in late June, however, led the Cabinet to ask to move forward the celebration to 19 July,[18] as a result of which the idea of the river pageant was abandoned and priority was given to the victory march—including last-minute plans for the erection of an ad hoc monument to be saluted by the military parade (Greenberg, 1989: 7–8; see also Homberger, 1976).[19]

In short, both Cabinet and Committee had their heart set on bringing a victory celebration to the British people, to the point of planning for it even amidst the incertitude of a not so clearly over war, and to adapt flexibly to international events to maximize participation and success. After all, patriotism was felt to be 'present in abundance in the street', and an explicit invitation to capitalize on that feeling by stressing 'the results of victory' was even made by Austen Chamberlain (Homberger, 1976: 1429).[20] On 19 July, then, a government-sanctioned 'peace celebration' took place in the form of a pompous victory march through central London culminating in Whitehall in front of a temporary Cenotaph unveiled on the day for the occasion (Greenberg, 1989: 9; see also Homberger, 1976: 1430). Locally arranged celebrations also took place on the same day throughout the country and empire (Homberger, 1976: 1430).

It is within this context of general unease and incertitude, with a British political elite focused on reinvigorating the public mood by actively trying to establish occasions and festivities to foster patriotism, that Armistice Day rites were born—and, I suggest, ought to be understood. The frenetic preparations for 19 July Peace Celebrations had probably captured the whole of the Cabinet's attention, to the effect of leaving the country 'unprepared' for 11 November. 'As late as 15 October 1919', Adrian Gregory (1994: 8) writes, 'there is evidence to suggest that the government had not considered the first anniversary of the Armistice as being a natural moment to consider the war and its human cost'. As Randall Stevenson (2020) notices, a letter

[18] For a more extensive and comprehensive chronology of the Committee's meetings and plans see Homberger (1976) and Greenberg (1989).

[19] Interestingly, there is a strong transnational dimension to this story. Indeed, both the ideas for victory celebrations and a victory march, and particularly the idea of erecting a war memorial, were deeply influenced by conversations between Lloyd George and other allied leaders. Both Homberger (1976) and Greenberg (1989) relate the story in detail.

[20] According to Homberger, Austen Chamberlain had been explicit in suggesting 'to work up a high patriotic feeling from the results of victory' (Homberger, 1976: 1429).

published in *The Times* on 4 November 1919 expressed appalment—if not honest preoccupation—that there were 'no signs of any official or public celebrations' scheduled for the first anniversary of Armistice Day, except for a 'big fancy dress ball for charity' to take place at the Royal Albert Hall (Howard, cited in Stevenson, 2020). Donald Howard, the letter's author, therefore called for additional and more inclusive forms of celebrations, arguing that 'we owe it to the dead as well as to our living selves *never to forget the victory that came*, so dearly bought but *so magnificently won*, on November 11, 1918' (Howard, 1919; my emphasis). And for Howard, doing so entailed establishing less upper-class focused celebratory practices to enable larger participation—such as a 'display of flags of the Allied nations' and a 'salute of guns' to be had at 11:00 a.m. on 11 November (ibid.).

Indeed, the 'big fancy dress ball' he referred to was one of many 'Victory Balls' being organized mainly for and by the wealthiest Brits at 'London's classier hotels' (Stevenson, 2020). These usually took the form of celebratory, charity dinner dances, held on the evening of Armistice Day (see Stevenson, 2020)—a practice which would firmly take hold until 1925, when a public debate would openly challenge and problematize it, finally leading to its demise (Gregory, 1994: 61–77). I will expand on 'Victory Balls' and the 1925 debate about them below, but for now, it is important to note the name of the practice, its general spirit, as well as the language surrounding it, at least in 1919 and the first part of the 1920s, are clearly focused on victory. And as emerges from Howard's letter above, so are the additional rites that he proposed—i.e., celebratory practices to mark victory in the Great War rather than reflective rites to commemorate the dead. As surprising as this may sound to the contemporary reader who is probably accustomed to the solemnity of today's Remembrance Sunday, the other commemoration practices that ended up on the agenda for the first anniversary of Armistice Day, including the Silence, had a similar victory spin to them.[21]

Given the scarcity of planned celebrations, and the general will to foster them,[22] it is perhaps unsurprising that when the proposal of a certain Sir Percy Fitzpatrick to institute a three-minute silence in celebration of the Armistice Day anniversary, on 11 November, reached the War Cabinet

[21] To put it like Mark Connelly (2002: 139), '[t]he Great War ended on 11 November 1918, the news was greeted with a day of wild, unrestrained rejoicing. How, then, did the anniversary of the Armistice take on a sombre, respectable form, observed with religious zeal and dedication?'

[22] Historian Adrian Gregory, an Armistice Day celebrations expert, also seems to espouse this position. He writes: 'In the immediate post war period, the danger of the example of the Russian Revolution spreading was a continuing preoccupation. The mood of returning servicemen was to some extent unknown, but there had been mutinies in bases in France caused by anger and uncertainty about demobilisation. … *If ever there was a time when rituals of integration and unity appeared to be required, it was the autumn of 1919*' (Gregory, 1994: 9–10; my emphasis).

on 4 November, it was immediately discussed, endorsed, and presented to the King for approval (Gregory, 1994: 8–12). Modelled on a South African tradition called the 'three-minute' pause, the ritual was meant to bring life to a halt for a short timespan with the aim of mobilizing collective emotions towards a specific purpose, namely to 'concentrate as one in thinking of those—the living and the dead—who had pledged and given themselves for all that we believe in' (Fitzpatrick, cited in Gregory, 1994: 9). Fitzpatrick's memorandum also specified: 'it is *not in mourning, but in greeting* that we should salute them on that day' (ibid.; my emphasis). The original silence, then, was not meant to be about mourning, but greeting and even celebration instead. At best, and only in hindsight, the practice can be thought of as potentially polysemous—if not (un)intentionally ambiguous—in nature, signifying at once for mourning *and* celebration, grief *and* relief, contemplation *and* homage. On this account, after all, it could unify the bereaved *and* the proud civil servant, the maimed veteran *and* the victorious general.

There is no doubt, however, that victory was central to the original ritual. The royal 'request for a national pause' that initiated the practice, published on all newspapers on 7 November, stated:

> Tuesday next, November 11, is the first anniversary of the Armistice, which stayed the world wide carnage of the four preceding years and *marked the victory* of Right and Freedom. I believe that my people in every part of the Empire fervently wish to perpetuate the memory of the *Great Deliverance*, and of those who have laid down their lives to achieve it. To afford an opportunity for the universal expression of this feeling, it is my desire and hope that at the hour when the Armistice came into force, the eleventh hour of the eleventh day of the eleventh month, there may be for the brief space of two minutes a complete suspension of all our normal activities …
>
> No elaborate organisation appears to be required. At a given signal, which can easily be arranged to suit the circumstances of each locality, I believe that we shall all gladly interrupt our business and pleasure, whatever it may be and unite in this simple service of Silence and Remembrance.
>
> **(King George V, cited in Gregory, 1994: 11; my emphasis)**[23]

[23] The link between the two-minute silence and victory is elaborated further, and in even more explicit terms, in the newspaper articles that promulgated the King's appeal. The *Daily Express*, for instance, reported: 'In quiet graves beyond the seas sleep a million British men who paid the price of *victory* … It is our duty to see that they did not die in vain, and for the accomplishment of that duty all classes must combine *as they did to win the war*, unselfishly and harmoniously. There must be a truce in domestic quarrels, an end to industrial strife. *We must all pull together lest the rewards of victory be thrown away*' (*Daily Express*, cited in Gregory, 1994: 11; my emphasis).

Firstly, the royal request references victory and the Great Deliverance explicitly. In fact, the King officially marks Armistice Day as a victory day, openly suggesting that 11 November 'marked the victory of Right and Freedom'. Among the climate of incertitude introduced above, it was now being made clear that the Great War was over *and* won. And that it was won on 11 November 1918—i.e., the request (retroactively) made Armistice Day coincide with the end of the war: not 28 June when the Treaty was signed, nor 19 July when 'Peace' was first celebrated, but 11 November. In other words, Armistice Day had a distinctively performative function: it materialized victory through its celebration. The request *timed* the war through victory.

Secondly, and relatedly, instructions are clear for the Silence to take place exactly 'at the eleventh hour of the eleventh day of the eleventh month', so as to 'symbolise perfectly the moment when the guns ceased firing on the Western front, signalling the end of the slaughter and the return of the prospect of life' (Gregory, 1994: 17–18). In short, not only was the Silence openly meant to commemorate a victory—i.e., 'perpetuate the memory of the Great Deliverance'—alongside the soldiers that made it possible, but it also sought to vividly re-enact (or, more precisely, perform) that victory, by endowing it with extremely precise temporal coordinates. Importantly, for Adrian Gregory (1994: 13–18) this specific temporal symbolism, a pause in, or a halt to, everyday life is among the key reasons for the Silence's immediate success—alongside the significant resonance with religious acts such as quiet praying, and 'its double nature as public and private commemoration' (Gregory, 1994: 18).

Lastly, in 1919 the Silence was already integrated with several ancillary commemorative practices, some of which were explicitly festive and celebratory in nature—and that, most importantly, would not be deemed inappropriate or irreconcilable with the spirit of Armistice Day until later. Beyond the Victory Balls introduced above, 'three cheers for victory' marked the end of the Silence in a south-west London factory (Gregory, 1994: 17), while a widely participated firework-lit march—accompanied by festive music and chanting—took place at 6 p.m. in Leeds (Gregory, 1994: 66). Additional associated rituals in 1919 included the congregation around (recently erected) local war memorials and the combination of the Silence with liturgical services such as Christian Masses (Gregory, 1994: 8–50). Notably, even sermons during Church Masses made explicit references to victory. Again in Leeds, for example, the Bishop of Ripon (cited in Gregory, 1994: 34) suggested that '[t]hese moments of silence might well be devoted to thoughts of thanksgiving, not only for the great victory vouchsafed us, but in that He had used us as his instruments in achieving that victory'.

Moreover, the two-minute pause had immediately proven its exceptional success: the Silence was extremely practical, a factor that resulted in large participation since its inception, and strikingly impressive. As evinced by the original royal request above, it was meant to be a 'simple service' that did not require 'elaborate organisation'. Most people marked the Silence simply by interrupting their mundane activities for the span of two minutes (see Gregory, 1994: 14). Spontaneous or semi-organized public gatherings of crowds were also relatively common.[24] In London, for instance, a large crowd gathered around the Cenotaph, where the royal family had sent a flower wreath in homage for display ahead of the Silence (Gregory, 1994: 15).[25] Public figures, such as Lloyd George, also joined to mark the two-minute pause (ibid.). Either way, the effect was magnificent, and it was clear the practice was there to stay: the whole country was choked with emotion as the pause succeeded in bringing life to a halt throughout the UK. Gregory (1994: 17) writes: 'The impact of the "Silence" had been enormous. It was immediately obvious that the general public had universally responded to the appeal and that furthermore it seemed to many people that the Silence (as it was soon to be called) ought to become a regular act of commemoration, occurring annually'.

## Reproducing victory as event: victory in the Great War from 1920 to today

Possibly on account of its unparalleled success, the Silence would be virtually undisputed from its inception until the present day. Other associated practices, language, and symbolism, however, would face a completely different fate. Victory Balls are a good case in point. As Gregory (1994: 61–77) explains, Victory Balls came under the public eye between October and November 1925 (ibid.) in what could be considered the first iteration of the 'victory in the Great War' debate introduced above.

With Victory Balls steadfastly taking place on Armistice Day for five consecutive years (Mansell, 2009: 449), dissenting opinions about the appropriateness of the practice and its characteristic festive attitude had been slowly

[24] According to Adrian Gregory (1994: 14) these usually took place at civic centres, places of worship, or around war monuments—most of which were still temporary in 1919. Curiously, the 1919 Cenotaph was also a temporary monument and would be erected permanently only in the following year on account of the success of the first Armistice Day ceremony (see Homberger, 1976; Greenberg, 1989; Connelly, 2002).

[25] This idea was so successful already in 1919 that plans were soon made for both erecting a permanent monument and making the practice of offering floral tributes part of a ritualized celebration in the following years (Gregory, 1994: 26).

mounting throughout the 1920s and emerged starkly in 1925 (Gregory, 1994: 67). In a letter to *The Times* dated 20 October 1925, for instance, veteran and vicar Canon H. R. L Sheppard noted that

> [A] fancy dress ball on a grand scale as tribute to the Great Deliverance which followed on the unspeakable agony of 1914–18 seem[ed] … not so much irreligious as indecent … Victory Balls of magnificence and their equivalent in thoughtless and ill-timed Armistice commemorations at hotels and restaurants should not be encouraged at least while this generation retains the heart-ache of a tender and thankful remembrance.
>
> **(Sheppard, cited in Gregory, 1994: 67)**

In his work on music and war commemoration, Mansell reports an anonymous letter published in the same outlet which is strikingly similar in tone:

> It is of course possible to make a plausible cause for anything—even for dancing on Armistice Day. Whether 'that victory feeling' is the real motive or only the excuse is for each to settle with his own conscience. The fact remains that most of us regard an Armistice Day ball as an outrage on decency. It does not prove callousness, it betokens a painful lack of imagination.
>
> **(*The Times*, cited in Mansell, 2009: 449)**

As Gregory (1994: 67–68) notices, such a position was not unanimous nor necessarily representative of a popular majority. In fact, defenders of Victory Balls—especially among war veterans—soon joined the debate, arguing 'that it was right to celebrate the achievement of the war, the victory won' (Gregory, 1994: 68). Writing again in *The Times*, for instance, a veteran sought to emphasize the importance of victory celebrations, and highlight that they had become controversial only recently, as Armistice Day 1919 was undoubtedly about victory:

> As one who is the only survivor of four brothers may I be allowed to say in regard to the Victory Ball that I have the feeling that the last thing that they would wish is that they should stand in the way of our enjoying ourselves. … Armistice Day has a double significance; it is a day of commemoration for the dead; but it was also a day of deliverance for the living. I do not see why both celebrations should not take place on the same day. *The first Armistice Day was a day of rejoicing*. But, the morning has now become, by usage, a time of remembrance. Why must that prevent the evening from being given over to the other aspect without any thought

> of disloyalty to the dead? *For November 11 was for all survivors quite tremendously the day of victory and deliverance.*
>
> **(*The Times*, cited in Gregory, 1994: 68–69; my emphasis)**

A *Daily Express* editorial also made a similar point: '[f]rom the earliest times—until this year, when an unwholesome and unnatural change is sought to be made—mankind has signalised its gratitude and joy for its successes in war, and its release from the travails which war brings, by prayer and by merrymaking. The two courses are not inconsistent' (*Daily Express*, cited in Mansell, 2009: 449). Another editorial from the same outlet reported that even General Herbert Belfield was a defender of victory celebrations. At best, for him, it was the recent mourning attitude that was out of place and in need of finding another day for its expression:

> 'I am entirely in favour of Armistice Day being regarded as an occasion for rejoicing,' said General Sir Herbert Belfield to a Daily Express representative yesterday. 'The solemn remembrance of the dead, should, of course, find a place in the day's celebrations, but I regard Armistice Day as a day for rejoicing over the victorious ending of the country's four year ordeal. Possibly the services in memory of the dead should be held, not on Armistice Day at all, but on, say, All Souls' Day'.
>
> **(*Daily Express*, cited in Gregory, 1994: 71)**

Other distinguished personalities and military generals also voiced their opinions, contributing to the debate's public resonance and visibility (see Gregory, 1994: 67, 70–74; Mansell, 2009: 449–450).[26] Eventually, and while several Victory Balls went ahead despite this heated debate, the Victory Ball at the Royal Albert Hall was cancelled, marking the epilogue of this public discussion (see Gregory, 1994: 73; Mansell, 2009: 449).[27]

While sparked by the practice of Victory Balls, it is clear from these excerpts that this debate raised broader concerns about the adequacy of certain commemorative attitudes and moral postures towards Armistice Day. Indeed, at least three different themes emerge starkly from the contributions above. Firstly, Victory Balls are epitomic of a festive 'victory feeling' that

[26] According to Adrian Gregory (1994: 70–77) a clear positioning of different outlets can be traced, with the *Daily Herald*, *The Times*, and the *Daily Mail* contesting the practice, and the *Daily Express* defending it. In addition, Gregory (1994: 74–75) provides evidence of additional opinion pieces on the same debate from *The Nation*, *Daily News*, and *St Martin's Review*, thereby giving a sense of the large resonance of the discussion.

[27] Importantly, a much more solemn remembrance service led by H. R. L. Sheppard (one of the debate's initiators) was held instead of the ball, which in turn was moved to 12 November to not give up on charity donations (see Gregory, 1994: 73–74).

has grown increasingly inadequate—or even 'indecent'—among (large) portions of Britons. Secondly, and relatedly, these individuals also deem such a 'feeling', or attitude, to be disrespectful towards both 'the dead' and the bereaved, arguing for a solemn posture (of 'mourning' and 'remembrance') that is, instead, far more appropriate. Lastly, defenders of this celebrative attitude point at its compatibility with 'remembrance' to highlight an emerging discrepancy between this 'new', solemn outlook and the original, joyful connotation of Armistice Day.

Far from being specific to Victory Balls, these themes will prove to be symptomatic of a general, emerging unease with the original meaning of Armistice Day commemorations. These reflections, in other words, would have a much broader remit than the practice that ignited their emergence: they express an attitude of discomfort with celebrating victory in the Great War which would consolidate throughout the following century, marking a shift away from celebration towards mourning and remembrance. According to Adrian Gregory (1994: 118–148), this radical shift takes place from the late 1920s through the 1930s, leading to nothing less than a 'transformation' in the meaning of Armistice Day—i.e., to the 'direct predecessor' of today's Remembrance Sunday. For him, a complex mixture of various factors account for this change in ethos, among which: a new sociopolitical climate marked by the collapse of 'the world economy' (Gregory, 1994: 118–119) and the election, in Britain, of a (pacifists-led) Labour government; the consolidation, in the wake of an increasingly truer new 'world-war menace', of the 'British Peace movement' (see Gregory, 1994: 149–183); or even 'the "War Books" publishing phenomenon' (Gregory, 1994: 118–119)—that is, the public affirmation of bestselling First World War memoirs such as *All Quiet on the Western Front*, *Goodbye to All That*, and *Memoirs of an Infantry Officer*.[28]

As he explains, the manifestation of this shift took different forms, such as the denouncement of Armistice Day as a militaristic event, or the call for a solemn attitude of the kind expressed in the Victory Balls debate above. Accordingly, traces of this change can be detected in a few, and perhaps minor, modifications to Armistice Day commemoration practices. Victory Balls, for instance, were officially dismissed in 1926 and they were substituted by 1927 by a British Legion organized 'Festival of Remembrance' (Gregory, 1994: 77–78) characterized by a respectful mourning attitude. Another subtle alteration is notable in the language surrounding well-established practices such as religious services or the Haig Poppy Appeal. In the first case, the

[28] For a detailed account and discussion of these and other factors see Gregory's (1994) chapters 4, 5, and 6.

idea of martyrdom—of 'men who had died for the *sins* of the world' (Gregory, 1994: 123; see also 184–211)—became central to sermons and orations, at the expenses of the victory-focused, 1920s ones discussed above.[29] As to the Poppy, it took on the contemporary association with remembrance—and support to the living[30]—exemplified by the Prince of Wales' (cited in Gregory, 1994: 109) exhortation, in 1930, '[t]o wear a Poppy ... to show that the wearer is not one to forget a common obligation [and to] pay for it generously—generously according to one's means is a proof of the sincerity of one's remembrance'. While by no means incompatible, such an exclusive focus on remembrance stands in contrast to the original, broader meaning of the Poppy as a token of 'remembrance, gratitude and sympathy' (Gregory, 1994: 102).[31]

But for the most part, Gregory (Gregory, 1994: 126; my emphasis) explains, 'the *form* of commemoration [of Armistice Day] changed little'—a statement that, arguably, still holds true today. In other words, this radical change in the *meaning* of Armistice Day was not reflected by a similarly stark alteration in commemoration practices. In fact, most rites and rituals remained virtually unchanged—a particularity that Gregory finds surprising. He writes:

> [B]ureaucratic conservatism meant that the ceremony at the Cenotaph was highly resistant to any form of innovation, so that even though there were many calls to change the nature of the ceremony to bring the symbolism into a closer relationship with changed thinking about the meaning of the war, the ceremony *was to remain surprisingly unchanged in form* throughout the 1930s.
>
> **(Gregory, 1994: 127; my emphasis)**

[29] As Gregory (1994: 118–148) notices, the rise of pacifism in Britain also played a key role here. With the shift away from victory and towards remembrance, Armistice Day rites risked becoming voided of significance: what did the Great War achieve? Why did all those men die? As a result, an increasing emphasis on the First World War as a sacrifice for peace emerged in the late 1920s and early 1930s, except being rapidly toned down as the shadow of the Second World War started to materialize.

[30] According to Gregory (1994: 93–117), the idea of supporting veterans through the funds raised with the Poppy Appeal was present since the practice's inception—but crucially restricted to maimed veterans returning from the Great War. This 'support' was progressively extended as to encompass Second World War veterans, and then all members of the military, returning from various conflicts—up to contemporary Poppy Appeals that are explicitly in support of ongoing war efforts (Kelly, 2023; see also Kelly, 2013; and Basham, 2016). Notably, it is also against this element of 'support for the living' that the alternative symbol of the White Poppy emerged in the 1930s (see Gregory, 1994: 149–183 for a detailed history).

[31] The original explanation of the Poppy's meaning is reported in full by Gregory (1994: 102) as appeared in the *Yorkshire Post* in anticipation of the first anniversary of the Poppy Appeal in 1922: 'This is Armistice Day, the day on which Germany acknowledged the defeat of her great and threatening armies. This is a day therefore of remembrance, gratitude and sympathy, a day on which we should all pay homage to the sign of the Red Poppy in token of our sentiments'. Importantly, and in line with the general celebratory attitude of the early 1920s, the explanation makes explicit references to Germany's defeat.

I want to conclude this section, however, by arguing that the existence of long-standing and enduring 'formal' continuities is, in fact, *un*surprising. Indeed, if we understand these continuities to be an expression—and a means for the reproduction—of a narrative that *performs* and *co-constitutes* the victory in the Great War, as I argued above, then it only makes sense that they persisted virtually unchanged. On display here is but a classic shift from active to passive timing—as Hom (2018a, 2020) would put it—that testifies to the success of this timing attempt. Moreover, endowing victory with (extremely) precise temporal coordinates—such as a specific date, or even time on that date—contributes to the reiteration of a (well-known) war trope in which the military, the social, and the political realms can be separated, and victory gained 'on the ground' alone.[32] Indeed, it is precisely through this narrational, timing process that victory ends up working as a timing indexical—i.e., obscuring a temporal process of synthesis and coordination by showing only its outcome. A closer look at the specific, formal continuities between Armistice Day and Remembrance Sunday should bring the point home.

As noticed above, despite its emergence as a fortuitous rite among the scarcity of planned celebratory rituals for the first anniversary of Armistice Day, the Silence was immediately successful, widely observed, emotionally resonant, and magnificent in effect. It is perhaps unsurprising, then, that the two-minute pause endured as the most notable continuity between Armistice Day celebrations in the 1920s, through the 1930s and the 1940s, up until the present day.[33] The Silence, I have argued, fulfilled the function of performing the victory in the Great War remarkably well: made to coincide with the official timing of the 'entry into force' of the Armistice of Compiègne at 11:00 a.m. on 11 November, it gave victory temporal substance by reproducing and universalizing the Western Front ceasefire. Notably, however, the rite had also been potentially polysemic in nature since its inception, arguably signifying for both victory *and* remembrance, collective public rejoicing *and* individual intimate reflection. In other words, the Silence already signified, at least partially, for remembrance—an undertone which started to be significantly amplified already in 1920 when the second Armistice Day celebration

[32] And, at best, according to this logic, the political pay-offs of victory exploited afterwards, as a direct consequence of a 'material outcome'—as I argued in Chapter 2.

[33] As Gregory (1994: 121; see also 122) notices, 'a re-examination of Armistice Day policy by members of the Labour cabinet' took place in 1929. Remarkably, the re-examination focused on the role of the armed forces and the importance of making Armistice Day a civilian rite, giving consideration, for instance, to the militaristic outlook of the Cenotaph ceremony. A public debate paralleled this re-examination, both in terms of time frame and contents (see Gregory, 1994: 123–126). In both instances the Silence was undisputed.

focused on the funeral of an 'Unknown Warrior of the Great War' (Gregory, 1994: 24; see also 24–28). Accordingly, the two-minute pause ritual did not need changing at all: the 'remembrance' connotation could just gain the upper hand over the victorious, celebratory one in the wake of the newfound meaning of Armistice Day.

The other two most notable continuities, namely the ceremony at the Cenotaph and the Poppy Appeal, underwent a similar process. With the advent of the radio and television, the former acquired increasing centrality within the general economy of Armistice Day/Remembrance Sunday (see Gregory, 1994: 138; see also 135–142). Live broadcast[34] on both media soon made the Cenotaph ceremony the Silence's backdrop par excellence in the whole country. The laying of wreaths also endured as a rite, corroborating an increasingly more solemn and mournful attitude enshrined in contemplative and less victory focused homilies and prayers (Gregory, 1994: 128; see also The Royal Family, n.d.). The sale of red poppies by the Royal British Legion also continued steadily, to the point that, as Gregory (1994: 111) explains, 'by the 1930s, the poppy appeal was a well-established part of British life' and 'had become in effect a social cliché'.[35] In short, poppies soon begin to *exemplify* remembrance, and together with more solemn and ritualized ceremony at the Cenotaph, they contributed to the solidification and normalization of the changed meaning of Armistice Day.

While all three practices could easily take on the contemplative and mournful overtones characteristic of the newfound focus on remembrance, they also did not break completely with their original, victory-focused features. As evidenced above, for example, both the Cenotaph and the Poppy were originally associated with victory—the former being quite explicitly *the* victory monument that the government erected on occasion of the peace celebrations and victory march on 19 July 1919 (see Homberger, 1976; Greenberg, 1989). When it comes to the Silence, this dynamic is even starker. In fact, the two-minute pause's victory-related symbolism was not abandoned at all. If anything, it slid quietly into the background, silently reproducing the fictional temporal parameters of the Great War victory year after year.

Firstly, radio and television broadcast led to the amplification and even sharper synchronization of the ritual, enabling the previously unthinkable

[34] The Silence and the Cenotaph service were first broadcast via radio in 1928 and on the television in 1937 (Gregory, 1994: 135, 142).

[35] Notably, a few practical and symbolic changes within the Poppy Appeal can be traced throughout the century—among which, for instance, the annual length of the sale (which expanded steadily), the material features of poppies on sale, as well as the designated recipients of the appeal's charity funds. For a focused discussion of the Poppy Appeal see Gregory (1994, especially chapter 3), Andrews (2015), and Basham (2016).

*precise* synchronization of the whole nation—or 'perfect co-ordination of time throughout the country' as Gregory (1994: 136) put it. 'It is an incredibly powerful idea that a whole nation should stop, in silence, at the same moment' Gregory (ibid.) writes, 'and radio, by capturing that immediacy, was the most effective way of transmitting that idea'.[36] As Marshall (2004: 40–41) explains, broadcasting enabled an 'interplay of sound and silence'—the 'chimes of Big Ben' now mark the beginning of the Silence everywhere in the country—that creates 'a momentary unity' by effectively disrupting a 'temporal continuity' *through* the Silence. In short, victory could now be knowingly and simultaneously re-enacted across the UK and its ruptural temporality actively performed to the effect of magnifying the pause's emotional resonance.

Secondly, the two-minute pause kept being observed at 11:00 a.m. on 11 November until 1945—except for the duration of the Second World War, when it was interrupted (see Gregory, 1994: 207)—and at the same time on Remembrance Sunday from 1946 onwards (Gregory, 1994: 212–224; more on this below). This is particularly significant considering the formal alteration of the title of the 'celebration'—which changed from Armistice Day to Remembrance Sunday in 1946—and the broadening of its scope as to encompass the remembrance of other conflicts' casualties. Speaking in his capacity as the Prime Minister, Clement Attlee informed parliament, in June 1946, that a decision was taken 'that the Sunday before the 11th of November, unless the 11th or 12th of November was itself a Sunday should be known as Remembrance Sunday, should be observed as a National Day of Remembrance for the wars of 1914–18 and 1939–1945 and should be associated with the National observance at the Cenotaph and the Two minutes Silence' (Attlee, cited in Gregory, 1994: 216). While ample consideration was given to alternative dates[37] in the institutional discussion leading to this decision, no alternative was found that could be as appealing as Armistice Day on account of the popular 'significance' it had acquired by 1945 (see Gregory, 1994: 219). According to Gregory (1994: 212–224), the shift to a Sunday was, in turn, actively sought by the British Legion and the Church to maximize participation in the Poppy Appeal and religious services.

[36] He further specifies: 'Clocks vary and getting the Silence to start and finish everywhere at precisely the same moment, so important in producing the psychological impact of total communion, would be impossible' (Gregory, 1994: 136).

[37] 'Dates tabled' Gregory (1994: 218) writes 'included, 8 May (VE Day), 6 June (D Day), 15 June (signing of the Magna Carta), 4 July (US independence) ... 14 August (signing of the Atlantic Charter), 15 August (VJ Day), 2 September (capitulation of Japan), 3 September (outbreak of War 1939), 15 September (Battle of Britain Day), 2 November (All Souls Day) and 11 November'.

In other words, not only did the 11:00 a.m. symbolism—and its continuous reproduction of victory's ruptural temporality—survive, unchallenged, a profound change in the meaning of Armistice Day, but it also endured undisputed when Armistice Day became Remembrance Sunday—a day of remembrance for the casualties suffered in both World Wars and, more recently, in subsequent conflicts, including ongoing ones (see Kelly, 2023). Here, the temporal nature of victory and the centrality of narratives to its co-constitution rings loudest. On the one hand, the performative (re)production of victory in the Great War persists unchallenged to the present day, testifying to the importance of narratives and their temporal dimension. Rather than leading to victory's total disappearance, the change in Armistice Day meaning adapted steadily to different contexts. Most importantly, its temporal symbolism almost took on 'a life of its own', ending up being silently detached from victory-fraught language and yet steadily reproduced. Victory in the Great War is not explicitly present in today's Remembrance Day rites. And yet, it is not precisely absent either. Rather, victory is quietly reproduced *in* and *through* time, in ritualistic performances of its temporal coordinates—or the concept's temporal structure—that keep the original victory narrative timidly alive.[38] On the other hand, a significant temporal discrepancy—or tension—is brought to the fore here: that victory narrative is just over one hundred years old. In other words, for the precise ascription of temporal finitude to take place, a continuing—and therefore temporally extended, or enduring—narrative is needed. To put it differently, it takes a very long time to make victory last one day—or two minutes.

## From mission accomplished to eternal glory: war videogames and the collapse of victory's temporalities

On 30 September 2022, the Imperial War Museum (IWM) in London launched 'War Games', the 'UK's first exhibition to explore what video games can tell us about conflict' (Imperial War Museum, n.d.). As the exhibition's webpage explains, videogames are 'one of today's most popular storytelling mediums' (ibid.), and 'Britain is a nation of gamers': '[in] 2021 more than half the adult population regularly played video games', registering a total expenditure over £7 billion on games (Imperial War Museum, 2022: 5). Accordingly, 'War Games' exposes its visitors to influential war-themed videogames

[38] To draw once again a parallel with Hom's (2018a, 2020) timing theory, a slow change from active to passive timing (see Hom, 2018a, 2020) can be detected in the narrational making of victory in the First World War in the UK.

alongside objects from the IWM collection, a military training simulator, and original video interviews with videogaming experts and developers to 'interrogate the thrill and tragedy of warfare in a game and its repercussions in the real world' (Imperial War Museum, n.d.). Indeed, videogames are not just popular and financially successful representations of armed conflicts and political violence. Rather, as both 'War Games' and a wealth of emerging scholarship suggest, videogaming *shapes* war in several, important ways.

Firstly, a very tight connection links the videogaming industry to military establishments worldwide. War-themed videogames are routinely used for both training (Stahl, 2006: 115–118; Leser and Sterrett, 2010; Robinson, 2019) and recruitment (Stahl, 2006: 122–125; Nichols, 2010; Nieborg, 2010) of military personnel. In turn, the implication of various national militaries in the development of videogames has also been shown to be profound, 'with the military providing consultancy services, access to military hardware' (Robinson, 2016: 255) and 'close collaboration' for the creation of 'story-lines' (Robinson, 2016: 259). Moreover, as historian J. C. Herz (cited in Huntemann and Payne, 2010: 4) explains, 'most of the technology that's now used in videogames had its origins in military research' and '[w]hen you trace back the patents, it's virtually impossible to find an arcade or console component that evolved in the absence of a Defense Department grant' (see Stahl, 2006: 118–122).[39] Recently, Aggie Hirst (2021: 483) argued that videogames are also central to post-deployment rehabilitation processes for military veterans, 'demonstrat[ing] that grassroots gaming groups promote community building, therapeutic relief, and suicide prevention'.

Secondly, videogames are central to contemporary representations of war—to the point of warranting a whole gaming-themed exhibition at the IMW. Today, according to Jarvis and Robinson (2021: 192), war-themed videogames are one of the key means through which war is made 'meaningful for publics around the world who are distanced from any encounter with "real" military combat' (see also Bos, 2018). On the one hand, scholars have shown that videogames are increasingly implicated in the propagation of militarist ideas and the militarization of society (see, for example, Stahl, 2006, 2010; Huntemann and Payne, 2010; Robinson, 2012, 2016). Nick Robinson (2015), for instance, found war-themed videogames to facilitate the popularization of the 'American exceptionalism' discourse, while Johan Höglund

[39] Accordingly, several authors have noticed that, nowadays, the boundaries between videogames' commercial and military applications are, scarily, increasingly blurred (see Stahl, 2006: 118–122; and Huntemann and Payne, 2010).

(2008) demonstrated that some of these games proliferate orientalist depictions of Western enemies (see also Hess, 2007; and Šisler, 2008). On the other hand, videogames can also ignite resistance to this 'militarisation of politics and society' (Robinson, 2012: 504), working as catalysts for political opposition to militarism, both from without (see Robinson, 2012: 513–519, 2019: 266–270) and from within (Hirst, 2021) military apparatuses.

Lastly, and relatedly, war-themed videogames are deeply entrenched with memory- and history-making practices. A study by Joel Penney (2010: 195–196), for example, found that players of Second World War videogames *Call of Duty* and *Medal of Honor* (both of which I analyse below) 'appreciated the games' serious historical content ... comparing the experience of playing the game to fighting a real war', concluding that 'rather than only providing escapist entertainment, the Second World War content connects players to historical reality in personally meaningful ways'. In other words, historical videogames are often approached by the player not as mere recreational tools, but as educational resources, especially on account of their 'persuasive language of historical authenticity' (Penney, 2010: 203). For Gish (2010: 168), while these same videogames 're-approriat[e] cultural and international history ... and manifestations of nationalized collective memory', they also 'have the potential to call into question fixed narrative histories that prescribe deterministic conceptions of the past' (on this point, see also Jarvis and Robinson, 2021). Either way, both public perceptions of videogames being educational, and scholarly reflections on their history- and memory-making roles suggest that the politics of videogames ought to be taken seriously.

Building on these ideas, in this last section, I argue that an additional, overlooked extrication of the relation between videogames and war is tightly linked to military victory. Videogames are yet another means through which victory is narrated and, therefore, produced. Just like the history textbooks and commemoration practices explored above, war-themed videogames are yet another victory practice that is aligned with hegemonic understandings of victory: they configure victory as a military, spatio-temporally circumscribed event endowed with causal power. However, videogames also represent another seldom explored realm within which victory is produced—one that is, again, seemingly peaceful, and far detached from the battlefield. By foregrounding time and temporality, I show that depictions of military victory in war-themed videogames contribute to the reproduction of timing operations that, while coordinating and synthesizing both the materialities of warfare and their narration, only show the material dimension of victory as their outcome.

## Playing the World Wars

One of the first, and arguably most famous, war-themed videogames of all time is *Medal of Honor* (1999)—a historical first-person shooter (FPS) game set in Europe during the Second World War. Conceived and directed by Steven Spielberg, *Medal of Honor* was released in 1999 amidst political controversies over its appropriateness (Edge Staff, 2015; see also Robinson, 2016). Nevertheless, the game quickly acquired huge popularity (Oldenburg, 2002), made over one billion dollars in sales since its release (Edge Staff, 2015), and continues to be a fundamental influence for videogames of the same genre to date (see Penney, 2010). Working closely with US Marine Officer Dale Dye—the same military advisor consulted for *Saving Private Ryan*—Spielberg sought to work on 'the game's sense of historical authenticity' and even embed 'miniature history lessons', creating a game that 'remains arguably the most educational FPS ever made' (Edge Staff, 2015; see also Penney, 2010). Since then, *Medal of Honor* has evolved into a hugely successful franchise counting seventeen titles, the majority of which retain the original historical focus on the Second World War, while others are set in more recent conflicts (Electronic Arts, n.d.). Unsurprisingly, several titles in the franchise have also garnered scholarly attention (see, for instance, Hess, 2007; Allison, 2010; Penney, 2010; Robinson, 2016, 2019). Importantly, however, the game's representation of victory—especially in its first title *Medal of Honor* (1999)—remains largely understudied.

In line with other early titles in the series (see Allison, 2010: 184), *Medal of Honor* (1999) is centred around an individual, Lieutenant Patterson of the US Office of Strategic Service (*Medal of Honor Gameplay, 2017*: 1:56–3:40). Assuming control of Patterson, the player embarks on seven, single-player covert missions against Nazi Germany that are key to the Allied success in the Second World War. Indeed, a short, opening video with varied historical footage of unspecified war-related scenes—including marching Nazi troops, artillery and war planes in action, speeches from Hitler, Mussolini, and Churchill, as well as images of the bombings of London—introduces the player to the Second World War, setting the start of game on 6 June 1944 (ibid.: 0:00–1:27). Aesthetically similar videos introduce and conclude each 'mission', providing some context and developing the game's narrative[40] alongside more frequent, written mission briefs that offer, instead,

[40] While some verbal interactions between characters is present in *Medal of Honor* (1999), these dialogues do not carry the narrative forward. Instead, they are mainly orders shouted from one soldier to the other, various imprecations, or other contextual exchanges that contribute to the game's ambience rather than to its narrative (see *Medal of Honor Gameplay*, 2017).

detailed instructions for each level—three to four per mission (*Medal of Honor Gameplay*, 2017).

While mission objectives are relatively varied—from rescuing an intelligence officer (mission one) and stolen art (mission six), to destroying a heavy artillery piece (mission two), a U-boat (mission three), and an atomic bomb prototype (mission seven), or invading German forts (mission four) and production facilities (mission five)—the actual gameplay unfolds analogously throughout all the levels: Patterson needs to kill Nazi enemies to complete a series of 'tactical' objectives—such as collecting objects, activating detonators, or accessing specific buildings—and unlock the next level (ibid.). Success, within this scheme, is the only viable outcome: failure simply result in the level being restarted until all objectives are attained so that the game can progress. In this sense, the entire gameplay and its very logic are premised on a narrow understanding of victory as the material outcome of conflicts.

First, *Medal of Honor* (1999) conceives of victory as made up of interlinked building blocks and times the war accordingly. Like the nested model of success from contemporary British doctrine discussed in the previous chapter, tactical successes in each level lead to 'operational' attainments (at the level of the mission), which in turn result in overarching victory at the end of the game—and bringing the game, and with it the Second World War, to a close. Within the set-up of *Medal of Honor*, in other words, levels function as self-contained tactical engagements that can be militarily won and, only cumulatively, result in a mission's (operational) success. Spatio-temporal containment emerges clearly, for instance, from the military briefs opening each level: besides clearly showing a date upfront, briefs are organized so that each mission unfolds in the span of one day or, at best, a few days (see *Medal of Honor Gameplay, 2017*). Moreover, the video clips at the end of each mission clearly mark their ending, displaying the successful results they made possible and recontextualizing their importance within the broader conflict. For example, in mission two—'Destroy mighty railgun Greta'—Patterson is first tasked with sneaking into a rail station, then gathering intel on a big piece of artillery named Greta, finding the actual artillery piece, and finally destroying it (ibid.: 24:53–44:24). The concluding video clip of the mission (ibid.: 44:25–45:01) shows historical footage of heavy artillery and other German lethal weapons being destroyed, closing on a Nazi flag being removed. With similar videos concluding each mission, the game successfully embeds each mission within a well-known historical narrative of the Second World War, up until the last mission—'The V2: Germany's Final Vengeance'—which is concluded by a last video titled 'Victory!' showing a shattered rocket and announcing the end of the war (ibid.: 44:25–45:01; 2:52:18–2:52:54).

Taken together, the spatio-temporal containment of the levels and the causal link that binds them reinforce the hegemonic understanding of victory as a martial event endowed with causal force problematized throughout this book.

Second, victory also emerges as essentially material in nature. The only actions leading to it are military operations that mainly entail fighting and killing. On one hand, the gameplay logic rewards killing, and killing alone, so that the true causal force moving the narrative of the game forward is military action on the ground—as stressed above. 'Performance evaluation' reports issued at the end of each level reinforce this idea (see, for instance, ibid.: 1:44:37). These reports provide the player with statistical information on their killing performance, displaying data such as number of killed enemies, shots fired and hits, alongside shot locations (head, torso, groin, legs, etc). Beyond magnifying the centrality of killing both to the game and, by extension, to the overarching war effort, the reports also form the basis for the attribution of medals that the player can collect throughout the game as a side objective. On the other hand, the idea of victory is also explicitly brought up quite frequently throughout the game. This is consistently done in the form of a poster depicting a soldier in uniform holding a rifle under the phrase 'On to victory!', displayed at the beginning of each mission's first level (see, for instance, ibid.: 1:46:37). Here, the poster can be read to articulate an aesthetic trope in which the gunned soldier—and by extension, killing—is taken to symbolize the key means through which victory is achieved.

Analogous victory logics are at play in several of the *Call of Duty* games, another major FPS franchise—arguably even more popular and financially successful than *Medal of Honor* (see Gish, 2010: 167).[41] The influence of *Medal of Honor* (1999) on *Call of Duty* (2003)—the first title in the franchise—is self-evident. Also set in Second World War Europe, *Call of Duty* (2003) offers a similar gaming experience: level after level, the player is tasked with fulfilling tactical objectives that mainly entail fighting and killing—such as destroying particular weapons, killing a certain number of enemies, or securing a specific area—with the aim of completing a series of military operations that cumulatively result in overarching victory (see *Call Of Duty 1 Gameplay*, 2016). Within the economy of the game, success is again the only viable outcome, with failure resulting in a level being restarted until all objectives are attained (ibid.). Moreover, levels and operations are also self-contained and spatio-temporally bracketed—the exact temporal and spatial

[41] As of May 2019, according to Nick Robinson (2019: 12; see also fn. 16), the *Call of Duty* franchise had 'combined sales of 300 million copies with total revenues of over $17 billion'.

coordinates (accurate to the hour) are provided at the beginning of each level, or mission—and victory references are similarly widespread, both in inspiring quotes appearing on screen throughout the game (see, for instance, ibid., 1:02:58) and dialogues between characters (see, for example, ibid.: 1:18:28; or 1:24:19). Just like in *Medal of Honor* (1999), the last mission in *Call of Duty* (2003), 'Storming the Reichstag' (ibid.: 2:16:49–2:22:50), collapses operational success with victory in the Second World War: the mission ends with the soldier controlled by the player waving a communist flag over the Reichstag and is followed by a short video clip of historical footage announcing the end of the War and the defeat of Nazi Germany (ibid.: 2:22:50–2:23:50).[42]

Unlike *Medal of Honor*, the first *Call of Duty* (2003)—as well as *Call of Duty 2* (2005) and *Call of Duty* 3 (2006)—is not specifically centred around one key, individual protagonist (see *Call Of Duty 1 Gameplay*, 2016; *Call Of Duty 2 Gameplay*, 2016; *Call of Duty 3 Gameplay*, 2020). Rather, the player takes control of three (*Call of Duty* and *Call of Duty 2*) to four (*Call of Duty 3*) soldiers enlisted in different Allied armies. While the campaigns played with each of these soldiers coexist and unfold in a shared space-time, these games put less emphasis on the narratives of the individuals controlled by the player, focusing instead on the convergence and contextualization of their stories within the broader historical trajectory of the Second World War—a key difference from the two remaining titles in the franchise's Second World War series, *Call of Duty: WWII* (2017) and *Call of Duty: Vanguard* (2021) (see *Call of Duty WW2 Gameplay*, 2017; *Call of Duty Vanguard Gameplay*, 2021). These later titles enact a fundamental shift—in line with other fantasy (as opposed to historical) games within the same franchise—towards foregrounding the protagonist's storyline and emphasizing their personal war experience of fighting the Second World War. Across all games in the Second World War series, however, *Call of Duty* portrays 'World War II as a closed event [to] establish[h] the presence of "real" history that is constantly referenced during the ensuing game play' while consistently 'positioning play as relevant to the outcome of the war itself' (Gish, 2010: 169–170). As a result, even titles that seek to home in on the war experience of specific soldiers rather than the overarching conflict they are part of, end up reinforcing

[42] Flags over the Reichstag's rooftop are a common victory trope across several titles in the *Call of Duty* franchise—indeed, they also appear in *Call of Duty: Finest Hour* (2004), *Call of Duty 2* (2005), *Call of Duty: Roads to Victory* (2007), and *Call of Duty: World at War*. *Call of Duty 3* (2006) ends with an even more explicit victory reference: a march under the Victory Arch in Paris. By contrast, *Call of Duty: WWII* (2017) and *Call of Duty: Vanguard* (2021) both end with scenes that bring closure to the individual biographies of their protagonists—in line with the franchise's narrational shift towards individual war experiences detailed below. For a comparison of all the games' endings in the franchise see *All Call of Duty Endings* (2022).

the idea that victory is a circumscribed event and causal force of military nature.

In other words, FPS games like *Medal of Honor* and *Call of Duty* are tied, by design, to a logic of victory as the natural outcome of martial events that obscures the narrational dimension of the notion. According to Gish (2010: 173), '[s]uch a conception of wartime history is clearly reductive, foregrounding singular acts of violence as the sole catalyst of military victory and the impetus for historical progression'. Just like the commemoration practices explored above, these videogames actively contribute to the reinforcement of a partial, if not inaccurate, understanding of victory by acting on its temporality. Narrating victory as the material outcome of war and a spatio-temporally bounded event endowed with causal power, videogames partake in the performance of victory while simultaneously effacing all the narrational efforts that enable it in the first place. To that end, victory is consistently represented as a martial event that comes at the end of war and brings it to closure. Importantly, this simplistic depiction of victory is constant across significantly different conflicts. *Medal of Honor* (2010) and *Medal of Honor: Warfighter* (2012) extend the same logic to the War on Terror (see Electronic Arts, n.d.), whereas the *Modern Warfare* and *Black Ops* series in the *Call of Duty* franchise apply it to fictitious modern conflicts and even entirely fictional wars against fantastical enemies, such as zombies. In other words, by embedding this logic of victory within the design of the game, FPS videogames universalize victory by reducing it to a simplified version of its temporal structure.

If the reproduction of such an understanding of victory can be attributed, at least partially, to the structural limits of game design, the idea is by no means exclusive to FPS games. Unlike mainstream shooter games like *Medal of Honor* and *Call of Duty*, *11:11 Memories Retold* (Bandai Namco, 2018) is not built around 'levels' to complete or 'objectives' to achieve, offering a completely different gaming experience. Created by some of the developers of *Valiant Hearts: The Great War* (see Kempshall, 2019), *11:11* is—just like *Valiant Hearts*—a narrative adventure game: a subgenre within war-themed videogames that is less centred on fighting and killing. Instead, videogames of this kind ask their players to undertake a set of (often mundane) actions to carry the narrative of the game forward. In short, *11:11* does not seek to entertain the player by having them engage in combat but by telling them a story. Nevertheless, an analogous understanding of victory characterizes the game.

*11:11* starts on 11 November 1916 in Toronto, Canada, where a young photographer called Harry is tasked with taking pictures of a uniformed man,

Major Barrett, who is about to go back to the western front after a short military leave (*11-11 Memories Retold Gameplay*, 2018: part I). Harry is in love with Julia, the daughter of his boss, and he cannot help but notice that Julia is fascinated by the man's uniform. Determined to become a military hero himself and conquer Julia's heart, Harry joins the Canadian army and travels with Major Barrett to Vimy Ridge. Meanwhile, in a Zeppelin factory in Germany, a technician named Kurt is also considering enlisting to further support his country's war effort. When the military unit of his son Max goes missing, Kurt's doubts are dispelled: he quits the job at the factory, joins the army and heads straight to Pas-de-Calais, his first destination (ibid.). After around an hour of gameplay, Kurt and Harry's storylines start intersecting. Both men are deployed in the same locality—only on opposite sides. When a tunnel collapses, Kurt and Harry are trapped underground and start collaborating—their supposed 'enmity' and linguistic barriers notwithstanding—to survive, get back to the surface, and ultimately return to their respective regiments (*11-11 Memories Retold Gameplay*, 2018: part II). Deciding whether to play as Kurt or Harry, the player of *11:11* follows the vicissitudes of both men for the remainder of the game, witnessing their friendship strengthen in spite of unfortunate events and war-imposed hardship (*11-11 Memories Retold Gameplay*, 2018: part III). Towards the end of the game, the player's agency is fully unleashed: different decisions in the two men's behaviours lead to six alternative endings—one that entails the death of both protagonists, one where they both survive, and others in which Kurt dies at the hands of different characters and Harry follows alternative life trajectories (see *11-11 Memories Retold All Endings*, 2018).

According to *11:11*'s producer George Rowe (cited in Smith, 2018)

> World War I games are normally a certain type of game where you're trying to shoot people in the head but for the common soldier, that's not what war is. They don't care about being there to shoot people. They care about whether they're going to get to see their family again. When they get home, are they going to be the same person? All this kind of stuff.

Both the choice of a narrative-heavy adventure game and the focus of its narrative arc clearly reflect this idea. *11:11* explicitly eschews traditional strategic tropes and the 'mainstream' emphasis on killings that characterizes popular military videogames, to instead focus on the 'lived reality' of war, as experienced by its two protagonists. As its producer put it, '[t]he desire was to tell a *real* personal account of war and then also to tell it from two different sides so you can get that different opinion on one conflict' (Tim, 2018).

The result is an overarching narrative that underlines well-known themes associated with war in general, and the Great War in particular, such as homesickness, camaraderie, death and mourning, humanness beyond enmity, and the 'inexpressibility' of war, its conditions, and horrors. These ideas emerge repeatedly throughout the game, both in the dialogues between characters, individual reflections of the protagonists, or in the epistolary conversations they maintain with their loved ones—and that the player has both access to and some agency over.[43] As 11:11's art director elucidates, 'war is something that happens to people rather than people happening to war. That was a nice appealing way to go in on the narrative' (Ttwheam, cited in Smith, 2018).

Unsurprisingly, given the focus on war experience and the explicit intent to communicate the First World War beyond 'strategy and killing' (Rowe, cited in Smith, 2018), the game does not have a victory focus comparable to the videogames analysed above: players are not asked to secure military objectives or fight incessantly to win the war, but to listen to, and develop, Harry's and Kurt's story instead. Military victory is not even prominent among the war tropes that characterize *11:11*, and the very idea of victory gets mentioned only a handful of times in over four hours of gameplay. This, however, stands in stark contrast with victory's temporal structure, which instead dominates the game's narrative arc. Indeed, *11:11* develops over a two years' time span—from 11 November 1916 to 11 November 1918—with each 'chapter' or 'episode' introduced by a countdown to peace. For instance, the game is opened by a black screen reciting 'November 1916, 2 years until peace' (*11-11 Memories Retold Gameplay*, 2018: 0:34), 'Part II' by the caption 'August 1917, 15 months until peace' (*11-11 Memories Retold Gameplay*, 2018: 1:21:34), and so on, until the last scenes of the game happening '1 hour before peace' and against a more frequent, and increasingly precise, countdown with minutes and seconds to 11:00 a.m. on 11 November 1918 when 'peace happens' and the game ends. In other words, while neither Harry's nor Kurt's stories are about victory per se, both unfold within a rigid time frame that reinforces the temporal boundedness of the war and the (supposedly precise) temporal coordinates of 'peace'.

While *11:11* explicitly seeks to spark reflection on the horrors of war by recentring soldiers and their humanity, it does so by uncritically assuming the conflict's spatio-temporal parameters and reinforcing a violent understanding of war as a temporal exception. *11:11* follows in the footsteps

[43] Several times during the game, the player has control over the content of the letters that Harry and Kurt send back home. For instance, the player can choose whether to soften the tone of the letter with omissions and lies in order not to exacerbate the recipients' preoccupation, or to be more truthful and blunter in describing the war experience (see England, 2018).

of commemorative practices like Remembrance Sunday analysed above: yet another narrative constructing victory as a vehemently material, martial event that efface the importance of victory's complementary social engineering. Nevertheless, on account of their popularity and political relevance, war-themed videogames and the narratives they promote cannot be approached as separate from the materialities of warfare—not when videogames are among the most recent recruitment tools in Western militaries, when troops are trained through them, and when broad audiences resort to videogaming to learn about war.

## Conclusion

> At the beginning, however, 'their' traits and 'ours', their acts and ours, had not yet been fully distributed. The disentangling of traits and acts—who inherited which—would be the narrative spoils of victory. *To the winners would go engineering rights to the story; to the losers, the traits and acts which would ensure that the story unfolded as the winners said it had.*
>
> **(Engelhardt, 2007: 17; my emphasis)**

> If history is made in the short run by the victors, historical gains in knowledge stem in the long run from the vanquished. ... *The experience of being vanquished contains an epistemological potential that transcends its cause*, especially when the vanquished are required to rewrite general history in conjunction with their own.
>
> **(Koselleck, 2002: 76–77; my emphasis)**

While both concerned with victory or victoriousness and their relations to epistemology, history-making, and knowledge production, the two quotes above reach almost antithetical conclusions. For Tom Engelhardt, to the winners goes the privilege of designing a story to which the vanquished can only adapt. On the contrary, Koselleck suggests that it is the defeated who bear the 'epistemological advantage' of twisting and adapting a narrative for the sake of their own future. Divergence of interpretations aside, on both accounts, stories—or victory narratives—acquires centre stage. As the last two chapters have shown, the materialities of warfare cannot be approached independently of the (victory) narrative(s) accompanying their interpretation. In fact, facts on the ground are co-constituted narrationally. Victory, in short, is not the outcome of conflicts, unless it is not constructed—or timed—as such. Rather, victory is a complex, social phenomenon intermingling war-making and its narration—an essentially temporal process that hinges on time and its production. And yet, the relationship between victory's material and narrational dimensions is rarely understood—let alone portrayed—in this fashion.

However, this chapter demonstrates that these insights are not specific to UK military institutions. I found these same logics and mechanisms to be at play in as disparate examples as history teaching in civilian schooling, commemoration rituals, and videogaming—three examples that are remarkedly non-military in nature. On the one hand, these victory practices, offer additional evidence that victory is co-constituted, narrationally and materially, within and without the battlefield. On the other hand, they also shed light on some additional features of the hegemonic understanding of victory at the heart of this book. First, the reduction of victory to the outcome of armed conflict is often underpinned by strong logics of linear causality. Not unlike history textbooks or war-themed videogames, the narrative articulated through Armistice Day's celebrations centred around identifying a specific moment as the end of war, and establishing an explicit, clear, and most importantly causal link between it and military operations. In fact, causal assumptions are even stronger outside the UK military—where doctrinarians acknowledge that narratives might change 'facts', the traditional victory story in the UK points in the opposite direction: it is only superior strategy, skilful soldiers, and a supportive public contributing to the war effort that can settle disputes or change history. Second, the social construction of victory as a temporally bracketed, time-altering event takes time—both in the sense of temporal operations of ordering, and an extended time frame. In the case of the First World War, for example, a two-minute victory took—and still takes—practices that are just over one hundred years old. A stark differentiation between wartime and peacetime, in this sense, sits at odds with victory practices that straddle these temporal phases precisely to constitute the very watershed event that sets them apart. As the next chapter will show, there is another sense in which victory practices meddle with these established ideas about victory: wartimes are often plural and not all of them necessarily end.

# 4

# War Bodies, Tattoos, and Stories

## Winning Wars That Do Not End

## Introduction

Among the ideas most frequently associated with victory is not only the end of war, but also victoriousness: a sense that victory can be someone's, or a sort of personal trait and achievement. Commanders, generals, and political leaders commonly exhort troops and nations to be victorious or remind them that they have been, *ex post facto*. Addressing the nation in May 1945, for example, Winston Churchill (in International Churchill Society, n.d.; see also BBC, n.d.) delivered a speech known as 'This is your victory' to simultaneously announce that the Second World War was over and that the victory *belonged* to the British people in front of him.

However, as a state or condition of superiority over a vanquished Other (see O'Driscoll, 2020a: 7), victoriousness does not presuppose that a given conflict has finished. This chapter [1] homes in on this idea of victory as individual quality and enquires into the temporal parameters it configures. To broaden the geographical scope of enquiry once more, I look at individual recounts of war experiences and their institutional re-mediations in the Anglosphere. The chapter argues that when the stories of those who have encountered war are heard closely, a plurality of wartimes comes to the surface, disputing the idea that victory necessarily brings war to an end, even when construed as a material event. In fact, in the war experiences that I attend to in this chapter, wartime does not necessarily have clean and clear-cut start and end points—not even when victory supposedly *belongs* to them. In these accounts, references to victory and other war endings range from scant to non-existent. War, in some of these stories, is akin to a *chronic* condition, whose very essence is a sense of temporal lingering that puts the idea of war's ending into question. And yet, institutional recounts of these

[1] An article version of this chapter has appeared in *International Political Sociology* (Palestrino, 2022b).

*Military Victory Beyond the Battlefield*. Mirko Palestrino, Oxford University Press. © Mirko Palestrino (2025).
DOI: 10.1093/9780198955818.003.0005

same war stories and experiences bring them back to hegemonic understandings of victory: they consistently unfold *chronologically*, silently reproducing a conventional notion of wartime that begins and ends at discrete points.

To make this argument, this chapter homes in on yet another victory practice, analysing four exhibitions of military tattoos. Tattoos, indeed, have recently become the latest genre of war art deployed by museums to make war tangible to their visitors. Since the first exhibitions of this kind was launched in the US in 2014, three additional ones were opened in the Anglosphere: US 2016, UK 2019, and Australia 2020. I show that both tattoos and museum exhibitions perform their timing functions by referring to material traces of warring and fighting. While veterans use tattoos to articulate and support their war stories, museums reread these same tattoos and their bearers' war experiences to explain organized violence to interested publics. Accordingly, unlike memoirs or diaries, exhibitions of military tattoos provide for a unique site in which individual war stories and their institutional re-mediations are simultaneously on display.

The chapter starts by contextualizing the proliferation of military tattoos' exhibitions within a longer tradition of war art, and a historically significant link between tattooing and the military. I then introduce a conceptual distinction between *chronic* and *chronological* timing outcomes—and attendant wartime experiences—to make sense of the temporal mismatches that arise from the individual war stories on display in these exhibitions and the institutional narratives about them. The chapter is concluded by the empirical analysis of tattoos and museum displays, and a discussion of the analytical and political implications of these narrational re-mediations.

## Representing war (experience): tattoos as war objects

On Veterans Day, 11 November 2014, a US-based virtual exhibit was launched 'that combines original video, photography, and audio to present the stories of military veterans in their own words' (War Ink, 2014; see also Dyvik and Welland, 2018). Arguably the first of its kind, 'War Ink' sought to narrate war by mobilizing a unique form of 'war art': military tattoos and the inked bodies of veterans. A huge success, 'War Ink' was soon followed by hybrid—that is virtual and physical—exhibitions: 'Symbol of Service' by the Illinois Library in 2016; 'Tribute Ink' by the Royal British Legion in 2019; and 'Ink in the Lines' by the Australian War Memorial in 2020.

While a relatively new genre, exhibitions of military tattoos follow in the footsteps of a long tradition of 'war art'[2] displays that grapple with representations of organized violence for the benefit of a wealth of audiences (see Brandon, 2007; Bourke, 2017). Indeed, the implication of museums and other 'powerful cultural institutions such as mainstream television and cinema' (Caso, 2020b: 55) in shaping 'visual regimes'—or 'mediascapes'—for the apprehension of war is well documented (e.g., Sylvester, 2009, 2019; McSorley, 2012; Scott, 2015; Welland, 2017; Caso, 2020a, 2020b). 'Particularly in late modern, post-conscription states where direct experience of war is limited', Kevin McSorley (2012: 51) explains, war art exhibitions and similar 'flagship cultural events are important practices through which wider public understandings of war ... are shaped and negotiated'.

Through the acquisition, selection, and display of war objects, museums produce and promote publicly available, influential war narratives (Scott, 2015; see also Sylvester, 2009, 2019). Always mediated and political, these narratives have often been described as memory- or myth-making exercises (see, for instance, Sylvester, 2019). Indeed, from commemorative and glorifying nationalist war myths (e.g., Caso, 2020a) to condemnatory anti-war narratives (see Brandon, 2007), war objects and other forms of war art have been used to support the most disparate political agendas (see Bourke, 2017). As archaeologist and anthropologist Nicholas J. Saunders (2004, 2017) explains, it is often the object itself, on account of its temporal qualities, that facilitates multiple, and even politically discrepant, re-mediations. For him, war objects 'are promiscuous, and, once made, can endure in the world far longer than their makers ... Whatever their original purpose, their sensual, emotional, psychological, practical and symbolic afterlives can be endlessly reconfigured, revalued and re-presented via new and previously unimaginable liaisons' (Saunders, 2017: 209).

In the context of a 'contemporary mediascape' for the apprehension of war that is 'decisively linked to the body and, in particular, to embodied experiences of soldiering' (McSorley, 2012: 48; see also Welland, 2017),[3] the recent proliferation of military tattoos exhibitions is perhaps logical. Indeed, as a specifically 'embodied and aesthetic expression of [personal] war experience' (Dyvik and Welland, 2018: 347), tattoos meet the canons

[2] The definition of war art is disputed (see, for instance, Brandon, 2007: 2–4; Bourke 2017). In this chapter, I use war art and war objects—broadly understood—to generally refer to artefacts which are employed by cultural institutions, such as museums, in support of their curatorial war narratives.

[3] Emerged against the backdrop of 'seemingly bloodless war[s] of machine-to-machine combat' (Welland, 2017: 526) in Iraq (1990) and Kosovo, the aesthetic regime surrounding the more recent conflicts in Afghanistan and Iraq instead recentred the soldiering body of a hypervisible 'liberal warrior', which the public is deemed to get to know intimately (Welland, 2017: 529–531; see also McSorley, 2012: 51–55).

of the contemporary mediascape exceptionally well. Moreover, a long history links tattoos to war and military personnel (Dyvik and Welland, 2018: 346; see also DeMello, 2000: 44–66). It is the colonial encounter of the Royal Navy with Tahiti, for instance, that originally brought tattoos to the West (DeMello, 2000: 45) and tattooing was widespread among Western serviceman even before its growth in popularity in wider society (DeMello, 2000: 48–50). In the US in the interwar period, DeMello (2000: 63) explains, '[t]he link between soldiers and sailors and tattooing was so strong ... that it was assumed a man with tattoos was serving in the armed forces or had been at one time'. Most importantly, tattoos still abound on the bodies of military personnel (Dyvik and Welland, 2018: 349)—patriotic symbols, military insignia, and commemorative motifs being particularly popular (DeMello, 2000: 50–51, 63–65).

This shared history notwithstanding, to date scholarly engagements with military tattoos belong almost exclusively to two disparate disciplines. First, a tradition of medical anthropology studies the health risks associated with tattooing (Gadd, 1992; Armstrong et al., 2000; Pearce, 2010) and the incidence of deviant behaviours among tattooed military personnel (Armstrong et al., 2000). Second, a historical literature explores the link between war, the military, and tattooing (e.g., Govenar, 1982; Balvay, 2008; see also Caplan, 2000; and DeMello, 2000: 44–66). Military tattoos are also sporadically mentioned in broader sociological and anthropological explorations of tattooing as a practice—for example, specific social dynamics associated with it (Atkinson, 2003; Kang and Jones, 2007; Sanders and Vail, 2008), meaning-making at both the individual and collective levels (DeMello, 2000; Kosut, 2000; Lemma, 2010), or the gendered (Braunberger, 2000; Mifflin, 2001) and cultural (Gell, 1993; Vaughan, 2007) underpinnings of tattooing. Engagements with military tattoos and tattooed soldiering bodies as war objects or pieces of war art, however, remain largely underdeveloped. As the only exception to this neglect, Dyvik and Welland's work (2018) draws attention to the communicative function of *both* tattoos *and* exhibitions, highlighting how the same tattoos and accompanying war experiences get caught up in different (re)curational exercises that result in complex, and even contradictory, war narratives.

The mobilization of tattoos, inked soldiering bodies and their attendant (war) experiences as war objects, however, simultaneously calls for an explicit focus on temporality. First, like the war objects in the work of Saunders mentioned above, tattoos' temporal dimension is also preponderant and complex. For art sociologist Mary Kosut (2000: 82), tattoos are 'communicative devices' with 'a distinctive narrative quality'. On account of their embodied and permanent nature, she explains, tattoos ground on one's body (continuously renegotiated and socially informed) individuals' narratives of the self, operating as physical temporal markers—or 'a historical reference point', as she puts it—aiding the emplotment of one's biography (Kosut, 2000: 96–98). In this sense, I suggest, tattoos work as a distinctive form of embodied timekeeping through which different personal experiences are not only 'documented' but also arranged, synthesized, and made to cohere for the sake of one's sense of biographical continuity. Tattoos, in other words, are (often explicit) temporal references concealing dynamic timing attempts and can be treated as timing indexicals. By doing so, in the analysis that follows I find that—as Kosut suggests—veterans make use of their (military) tattoos to articulate their war experience, anchor it to their bodies, and reconcile it with their narratives of the self. Importantly, however, I also find that exhibitions' curators similarly resort to tattoos and accompanying war experiences to ground and support—that is to *time*—their war narratives.

Second, indeed, curatorial arrangements of tattoos, inked soldiering bodies, and accompanying war experience (i.e., their use as war objects) should also be understood as—contextual, and sometimes competing—timing operations. As Synne Dyvik and Julia Welland (2018: 346–347) argued with reference to 'War Ink', '[n]ot unlike the work of a museum curator, the web archive selects, frames, and displays individual veteran tattoo narratives within a wider schema of militarized aesthetics', thereby seeking to 'curat[e] and interpre[t] the meanings of these experiences for the wider public'. By reading both tattoos and tattooed bodies as timing indexicals, the temporal nature of curatorial practices becomes apparent. The 'reconfigur[ation], revalu[ation] and re-present[ation of these objects] via new and previously unimaginable liaisons' (Saunders, 2017: 209) is but a timing operation, performed by a different timing actor, to serve a different 'will to time' (Hom, 2018a: 73). In these exhibitions' war narratives, that is, war experiences are not reconciled to the biography of their protagonists but to the narrative frames—such as heroism, patriotism, and national pride—that best suit the exhibitions' political commitments.

Indeed, as Hom (2018a: 75) explains, 'multiple timing practices' might well coexist, to the effect that the same timing indexical can sit at the centre of different timing efforts. However, when the same tattoo ends up supporting both an individual's narrative of the self *and* the war narrative(s) promoted by the exhibitions, narrational discrepancies come to the surface. Here, the issue of war endings is a case in point. While different understandings of wartime emerge from veterans' personal accounts of war, exhibitions are univocal in portraying war as a discrete, temporally finite event. As I show in the analysis below, some veterans locate the beginnings and endings of their war experience at diverging temporal junctures. Some others articulate a war experience that has no clear starting and ending points in the first place. Far from being a mere nuance, these embodied temporalities of war experience and their attendant understandings of wartime are qualitatively distinct. This difference can be understood in terms of the distinction between *chronic* and *chronological* temporalities.

Although semantically interlinked, the time-related qualifier 'chronological' and its linguistic predecessor 'chronic' (see Freeman, 2019: 124–157) are conceptually different. Born as a neutral, general term to designate something as related to, or about, time, 'chronic' slowly came to be associated—during the eighteenth century—with states of persistent illness and other protracted medical issues (Freeman, 2019: 125). As a result, its meaning was crucially altered, gradually acquiring both a negative connotation and references to unfinished, lingering temporal states. Unlike 'chronology', which designates orderly sequences of events, 'chronicity', Freeman (ibid.) explains, 'correlates with a certain shapelessness in time, and chronic conditions seem to belie narrative altogether'. In chronic conditions, she further specifies, 'what is "continuous" and "prolonged" is as likely to be the past as the present, and the two cannot be fully distinguished' (Freeman, 2019: 145)—a feature that makes references to beginnings and endings altogether inadequate. Importantly, though, the original semantic association of the concept with time and temporality is retained, enabling subtle narrational slippages between the two terms.

Expanding on Hom's timing theory (2018a, 2020), I suggest that 'chronic' identifies the result of a timing attempt in which the relation between continua of change is in place, but also crucially complex and cloudy—as if the ordering was, at some point, somehow stuck. Unlike the orderly sequences of phenomena characterizing its *chronological* counterpart, a *chronic* outcome identifies messy relations between change continua. First, *chronological* and *chronic* designate two qualitatively different embodied temporalities of war experience—orderly recounts of war experience that might (but do not necessarily) diverge from the chronology of a given conflict in the first case, and lived experiences challenging war's temporal finitude *tout court* in the second. Moreover, and second, distinguishing between *chronological* and *chronic* timing outcomes simultaneously sheds light on the politics of curatorial re-mediations of war experience—such as those instantiated by the exhibitions of military tattoos under analysis—which portray war as a temporally circumscribed event.

Attending closely to the individual and institutional narratives on display in these exhibitions, the following two sections home in on the timing operations that tattoos harbour, differentiating between those performed by veterans' and by the exhibitions. Temporally (un)bracketing the war experience by inscribing (bits of) it onto the human flesh, tattoos illuminate war's complexity, its multifaceted heterotemporality and the ephemerality of its (temporal) contours—thereby shedding some valuable light on the concept of victory and its temporal politics.

## Inscribing war, eschewing finitude: the wartimes on inked veterans

Even at quick glance, the number of direct references to time and temporality found across the four exhibitions under analysis is striking. One of the most recurrent temporal themes on veterans' inked bodies is dates—depicted either as self-standing subjects or as elements in broader displays. Inscribed on the lower back of US Landing Support Specialist John Miles are ten dates, listed, one below the other, in chronological order (Illinois Library, 2016a). These, John explains, 'are the dates of ten of the Angel Ceremonies I participated in' when deployed to Afghanistan in 2010, 'a reminder I should live up to my potential on a daily basis out of respect for them, because they don't have a chance to anymore' (Miles, in ibid.). ID tags reporting names, surnames and *dates of decease* of six comrades lost in combat hang from a chain held by praying hands on the back of Paul Glazebrook, above an excerpt from the famous Binyon poem 'For the fallen' (Tribute Ink, 2019a). Other recurrent dates found on veterans bodies include deployment dates (see, for instance, Miles in Australian War Memorial, 2020a; 2020b) and enlistment and discharge dates (e.g., Adam, in Australian War Memorial, 2020c).

Perhaps even more popular among time-related military tattoos are varied exhortations not to forget. The motto 'fallen but not forgotten' is inscribed on the skin of two of the twenty-four veterans from the War Ink project, namely US Combat Engineer Johnathan Snyder (in War Ink, 2014: Chapter 2) and US Armor Division Officer Russell Toll—in this last case, in the Latin version 'abitus sed non oblitus' (ibid.). In a similar vein, US Army Infantry Sergeant

Garret T. Anderson complemented the tattoo of a winged battlefield cross[4] on his left leg with the phrase 'never forget' (Illinois Library, 2016b). This same expression, 'never forget', is also part of Australian Army veteran Chris's display, an ensemble of poppies, dates, mottos, and a soldier silhouette, realized to commemorate his participation to the centenary of Anzac at Gallipoli (Chris, in Australian War Memorial, 2020d). Related slogans, such as 'lest we forget' (e.g., Adam, in Australian War Memorial, 2020c; Shawn, in Australian War Memorial, 2020a) and 'we will remember them' (Newton, in Australian War Memorial, 2020e) are also quite popular.

Remarks about the autobiographical, or timing, function of their tattoos are similarly widespread among veterans. Talking about the purposes of his body art, US Navy Corpsman Joel Booth explains how for him 'tattoos are meant for *documenting* ... life experiences' (Booth, in *Doc*, 2014, my emphasis). On a similar note, US Army medic William Glazier suggests that he 'just did not want to forget anything that happened ... So a tattoo has that permanence where ... it will always be there to remind you' (Glazier, in *This is for Pape, 2014*). US Air Force mechanic Heather Hayes conceives of her tattoos as 'a journey', specifying that 'I don't think there's a single tattoo I will ever regret, even if it's a phase that I grew out of, 'cause it represents something that I was going through at that time' (Hayes, in *Suicide Butterflies*, 2014). Paul, a veteran from the 'Ink in the Lines' exhibition, explains that for him 'some tattoos are good decisions and some are bad when you look back, but that's just a *piece of time*' (Paul, in *Commemoration—Ink in the Lines*, 2020, my emphasis). Albeit not explored in any detail, the timekeeping—or autobiographical—role of tattoos is also echoed by the War Ink creators, who suggest tattoos are 'very powerful *records* of moments, people, times' (Deitch, in *Interview with the Creators of War Ink, 2014*).

Echoing this understanding of tattoos as records and documents—or, more generally, as time-harbouring devices—US Combat Medic Brent Blackwell, one of the seven veterans interviewed for the Symbols of Service Exhibit organized by the Illinois University Library (2016c), reported getting military tattoos because he 'wanted to take a part of [his] life and *document* it' (Blackwell, in Illinois Library, 2016d, my emphasis). For US Infantry Sergeant Caleb Carlson tattoos 'tell the stories [he] want[s] to convey' (Carlson, in Illinois Library, 2016e). Similarly, Craig Daniell, a UK Senior Aircraftman whose body art is part of the Tribute Ink online exhibition, thinks that a 'tattoo is a visual memory of [his] time in service' (Daniell, in Tribute Ink,

[4] A battlefield cross—or battle cross—is a temporary commemorative symbol to mourn the death of a soldier. Usually, battlefield crosses are made of the deceased soldier's rifle, boots, and helmet. For a brief account of its history, see Kirkpatrick (2018).

2019b) and 'a story that if people do ask [he] can explain' (Daniell, in *Tribute Ink—Craig Daniell*, 2021). Talking to the curators of the Australian War Memorial Online Exhibition 'The Lines of Ink', veteran Kevin explains that 'we [soldiers] get a tattoo to remember, and commemorate, our service, our mates, our careers' (Kevin, in *Mateship—Ink in the Lines*, 2020)—an idea David expands upon extensively:

> I wanted a story. So when people do ask, 'cos they do get interested, I wanted to be able to give them a story of what I've done, places I've been and things like that. To me it's a reminder every day. You don't need to look thru photos, you can just look at your arm and go, 'Ah, I remember that'.
>
> **(David in *Identity and Belonging—Ink in the Lines*, 2020)**

For these veterans, in other words, tattooing has a timing function—that is, it enables ordering, synthesizing and stabilizing continua of change into a coherent bibliographical narrative. Indeed, the temporal traces registered in, through, and by tattoos are not necessarily exclusively past-oriented. Instead, veterans seem to employ body art as a proper sense-making device that enables (re)orientations of the arrow of time in different directions. Recollecting some of the specific events that he wanted to ink on his body, Chris Markowski, a Marine who served in Iraq, explains he sees tattoos as 'sort of like, rite of passage' (Markowski, in *The Unwanted, 2014*). For Chris, the military provides people with 'the ability to make [their] own future' (ibid.) and his tattoos 'speak' to that. Hinting to future-making more boldly, Jeff Slater, an army small weapons expert, suggests that in his understanding of the practice, getting a tattoo is 'like putting your personal *goals* on your body' (Slater, in *Bombs to Butterflies*, 2014, my emphasis)—a constant reminder that he is 'trying to develop into something better than [he is]'.

Among this multiplicity of time-centred tattoos and narratives, three pieces stand out for the way in which they exemplify tattoos' timing function while simultaneously foregrounding their bearers' relation with—or understanding of—the idea of wartime. Inked on UK Royal Air Force (RAF) Senior Aircraftman Craig Daniell's right leg is a full-sleeve tattoo recounting his 'time in service'—that is, two deployments to Afghanistan in 2011 and 2013 (Tribute Ink, 2019b). Several elements compose the massive design, among which: a lion face, a medic, some poppies, a cruciform shape inscribed with the name of a lost comrade—Ryan Tomlin—and the date of his death, a parchment or papyrus-scroll-shaped inscription, stating 'stay patient and trust your journey, it's a matter of time', a post-traumatic stress disorder (PTSD) ribbon, and a long stairway conducting a man in military uniform to a huge clock (ibid.). Each of these elements plays a specific role in narrating Craig's service. The medic, for instance, represents his first experience in a Medical Evacuation Response Team; poppies and the cruciform shape commemorate lost comrades, while the soldiers approaching the clock signifies 'his belief that his time was up when he was discharged from the RAF' (ibid.; see also *Tribute Ink—Craig Daniell*, 2021).

Craig got the sleeve after being medically discharged from service: '[i]t wasn't until I was injured myself that I stopped and had time to think and my brain went into overload ... After diagnosis for PTSD, I realised I had to take my time and really process everything and reassess myself after injury' (Tribute Ink, 2019b). The tattoo, he adds, 'was a way of getting it out' (*Tribute Ink—Craig Daniell*, 2021). He explains: 'this is a story that I don't always want to tell people anyway ... I don't go up to people and say I was MERT [Medical Emergency Response Team] before, I picked up casualties, because it's not a thing we always say, so this is like a story so if people do ask I can explain the situation instead' (ibid.). For Craig, then, his tattoo is a way to communicate *his* story, a story he only got to—and 'wrote' on his leg—after recovering; but also a story that is about *his* war experience, one over which he explicitly seeks to assert control. Not only is this story simultaneously about war *and* Craig—that is, a phenomenon timed by Craig with his own autobiographical arc as its timing standard. Most importantly, his tattoos do not explicitly relate the end of the story, allowing for multiple and even continuous re-articulations. While time figures prominently, it has blurry and continuously (re)negotiated boundaries. In fact, war, as represented on Craig's body, has no clear temporal contours. Wartime, as depicted on Craig's body art, extends beyond his 'tours': temporal references *in* and *on* the sleeve span from pre- to post-deployment without really ordering events in a fixed sequence. The chronic temporality of his war experience also arises from the multiplicity of incongruent temporal markers that populates and cohabits the full-sleeve tattoo such as the (claims to) eternity of memorials such as poppies and crucifixes and the atemporality of a symbol such as the PTSD ribbon.

In an attempt to instead temporally bracket his war experience, US Army Combat Engineer Thomas Lancaster has opted for the Arabic translation of the phrase 'lost time' on his right arm (War Ink, 2014: Chapter 4). Accompanying his tattoo, in the War Ink exhibition, is a quote from his interview: 'We had 15 months just lost out of our lives. And by that time, the harsh reality sinks in ... Life goes on without you. And all our families continue on without you. And it's like a dream' (ibid.). Both tattoo and quote suggest that Thomas understands wartime to be a (lost) temporal parenthesis. In contrast to other narratives emerging from the archives, Thomas's war experience has a clear duration. However, *his* war in Iraq begins and ends with his deployment. Indeed, the chronological temporality of Thomas's war experience emerges strongly from the combination of tattoo and quote: inscribed within his autobiography, Iraq is juxtaposed to a non-war-torn storyline and (dis)qualified as, literally, a loss of time.

On the left hip of US Navy Corpsman Joel Booth is a vignette of the grim reaper holding a baby by the right leg—the leg Joel lost during an improvised explosive device (IED) detonation—and dragging him to the grave, where a tombstone engraved with Joel's date of birth and deployment awaits him (see *Doc*, 2014). Other suggestive tattoos populate the veteran's body. In particular, on his right shoulder and right-hand side of his back is a huge skull wearing a chief headdress—the emblem of Joel's unit—with a headband stating 'Make Peace or Die'—the unit's motto—next to the phrase 'only the dead have seen the end of war' (ibid.; see also War Ink, 2014: Chapter 2). While Joel does not expand on the meaning of that specific phrase in his interview, nor provide a specific reason for getting that particular sentence inked, the *temporal infinitude* of wartime it expresses is self-evident. War cannot be over. In fact, 'only the dead have seen the end' of it. The temporal reference in the quotation (i.e., the idea of the end) ascribes the tattoo to the domain of temporality, making it a timing indexical worth studying. At the same time, the quotation attributes the capacity to end war only to the lifeless, thereby denying all living beings the chance of experiencing wartime's temporal finitude. As a result, the veteran is left with a never-ending war experience that he does not wish to (or perhaps cannot) forget, but from which he does want to move on (see *Doc*, 2014)—that is, in a chronic temporality he cannot ignore nor escape.

Notably, the temporal understanding of the war experience encapsulated in this last tattoo seems to be popular among veterans. Different anthropological accounts resonate with this idea of war as 'unfinished business'. In *Making War at Ford Hood*, Kenneth T. MacLeish (2013: 7–8, my emphasis), suggests that 'attending closely to soldiers' experiences reveals the ways that war is not at all an exceptional condition, that soldiers ... are *always already* in the middle of it and its *unfinished* present'. According to Zoë H. Wool (2015: 145), this is attributable to the 'deep, ontological transformation' soldiers undergo because of the war experience. In her ethnography of life at Walter Reed Army Medical Center, a military facility for the rehabilitation of injured soldiers in Washington D.C., Wool (2015: 131–155) finds veterans' 'movement' to be viscerally altered by war. Defining movement as 'an attempt to capture carnal and emplaced experience, flesh and the socially emplotted ways it is seen and felt, proprioception and those other senses of sight, sound, touch, and taste through which a body and a space enact a meaningful, sensible articulation', Wool (2015: 133) hints at a deep, and most importantly, permanent alteration of soldiers' being. For veterans, in other words, the war experience is *chronic* or never over, as war (out)lives *in* and *through* them and their bodies, beyond combat and deployments. In short, war produces soldiers as (radically changed) social subjects.

Interestingly, and the abundance of temporal traces notwithstanding, no references to war endings are found among the body art on display. The only partial exception to this absence is a reference to victory and defeat (i.e., war's classic temporal markers) in a quote from Theodore Roosevelt, inked on the right forearm of US Army Animal Care Specialist Tracey Cooper-Harris (*Dare Mighty Things*, 2014):

> Far better it is to dare mighty things, to win glorious triumphs, even though checkered by failure, than to take rank with those poor spirits who neither enjoy much nor suffer much, because they live in the gray twilight that knows not victory nor defeat.
>
> **(War Ink, 2014: Chapter 4)**

Tracey sums up her understanding of the quote as meaning: 'You know what? Just go out there and do it!' (*Dare Mighty Things*, 2014). Unsurprisingly, then, Tracey links her tattoo to both her experience in the military and her role as activist afterwards, a time spent fighting the 'Don't Ask, Don't Tell!' policy in an attempt—later proved successful—to change the law (ibid.). In other words, Tracey's body art is yet another instance of autobiographical work. Overall, she suggests, her tattoo testifies to the importance of veterans' labour—one that is not restrained to the time of service, but expands beyond that, as veterans 'are also trying to do the things that help our country, our nation, and our society as a whole, do bigger and better things' (ibid.). While the quote resorts fervidly to both the concepts of victory and defeat, Tracey seems not to understand them as necessarily war-related, nor as temporal markers. Instead, in her interpretation, both terms are deployed metaphorically, as periphrasis for success and failure, and their meaning is explicitly extended to non-combat situations. Most importantly, the contextual information that Tracey provides makes clear how her tattoo does not really connect victory, nor defeat, to war endings. If anything, the tattooed quote works 'across temporalities'—or in accordance with Tracey's autobiography—relating Tracey's story both during and after deployment.

Importantly, this is not to suggest that interpretations of war as temporally unbounded in the lived experience of soldiers are a univocal position emerging from these body art archives. Instead, I wish to draw attention to the temporal complexity and multiplicity arising from the engagement of these military tattoos. While time and temporality figure prominently on the inked bodies of veterans, there is no unitary narrative—nor a clean, clear-cut understanding of wartime and the war experience—connecting their body art. Instead, as the above examples sought to emphasize, for some, deployment is a convoluted event one could try to grasp a posteriori, whose temporal parameters are in constant negotiation; for others, it is a bracketed period within their own longer temporal trajectory; for some others yet, the time of deployment is utterly unbounded, to the point that it is actually never over. However, as I detail below, this heterogeneity clashes outright with the exhibitions' univocal re-mediation of these same war experiences through the discourse of a temporally bracketed wartime.

## (Re)timing war bodies: chronic wartime and chronological renderings

That the general register of these exhibitions conveys and reifies the idea of a temporary, exceptional, and past-confined wartime is immediately visible to the virtual visitor. The first slide of the 'War Ink' website, for instance, explains that veterans 'struggle to process their wartime experience, recover from profound losses, and reintegrate into civilian life' (War Ink, 2014, my emphasis)—that is, explicitly marks wartime as a past, exceptional period of time veterans seek to recover from and juxtaposes it to peacetime. Similarly, upon entering the 'Tribute Ink' website (Tribute Ink, 2019c), the virtual visitor is invited to discover 'how tattoos have become central to Remembrance and commemoration in our [i.e., the UK] Armed Forces community'. Accordingly, throughout the different webpages composing the exhibition, references to war and fighting are provided in the past tense, and tattoos explicitly conceived as commemorative (i.e., past-oriented), memory-marking tools—an idea further reinforced by the thematic areas of the exhibitions, three out of four linked to remembrance and memorialization ('rethinking remembrance', 'remembering the fallen', 'marking the memories'). On a similar note, the 'Ink in the Lines' opening webpage immediately explains that veterans 'share a common purpose in getting inked: to remember' (Australian War Memorial, 2020a).

Mobilized in support of these narratives, tattoos and inked soldiering bodies keep serving a timing function. Unlike veterans' stories above, however, these narratives do not adopt an individual's biographical arc as their timing standard. As a result, for instance, body art pieces interpreted by their bearer as personal goals or 'reminder[s to] live up to [one's] potential' (Illinois Library, 2016a) simultaneously come to signify symbols of 'pride in service, patriotism ... and memorials' (Illinois Library, 2016c). Submitting the exhibitions' curational register to the same type of analysis, timing theory also illuminates the temporal politics of curatorship—as well as any other re-mediation process of these war experiences. As suggested above, a given timing indexical can signify, or hint at, different timing attempts (see Hom, 2018a), working as an entry point into multiple, overlapping narratives. Importantly, the outcome of the concurrent and overlapping timing processes these clues might conceal—such as narrational re-mediations—can also be classified as chronic or chronological. As temporal discrepancies among narratives arise, specific instances are found in which the alteration of the timing standard simultaneously results in the replacement of chronic outcomes with chronological ones. In turn, the result is a reification of a monolithic understanding of wartime as momentary and finite. In short, treating tattoos and inked soldiering bodies as war objects, some of these exhibitions' war narratives re-mediate the chronic references found above in chronological terms.

In 'Tribute Ink', the inked bodies and related stories of nine different veterans are grouped thematically. A dedicated webpage introduces each theme, providing some rationale and contextualization, as well as hyperlinks to the individual pages of two or three veterans—depending on the theme. In turn, veteran-specific pages provide a combination of textual material—direct quotes from interviews and some other contextualization—pictures of the tattoos, other pictures of the veteran, and, for some of the subjects, a brief chunk of a video interview. Both the text excerpts and video interviews are resonant with the broader theme to which the veteran-specific webpage is associated. As for the visual displays, it would seem they are designed to corroborate the narrative within which they are collocated. For instance, UK Royal Navy Logistician Danielle Cummings' webpage revolves around the themes of the family and the navy, in line with her tattoos—anchors and swallows (Tribute Ink, 2019d). Besides the pictures of the tattoos, then, the rest of the visual records on the webpage depict Danielle on board a frigate, while the video is set in the very familial environment of Danielle's home (*Tribute Ink—Danielle Cummings*, 2021).

Interestingly enough, for a series of veterans—namely Craig Daniell (Tribute Ink, 2019b), Matt Tomlinson (Tribute Ink, 2019e), and Paul Glazebrook (Tribute Ink, 2019a)—whose tattoos (or attendant narratives) are specifically related to the idea of remembrance and memorialization, the aesthetics of the non-tattoo-specific pictures on their webpage follow the same, especially distinctive, pattern. Each of these veterans, indeed, is photographed on a memorial site and their tattoos and/or tattooed bodies are juxtaposed to—or put in conversation with—the monument. Neither the text nor any other element in their dedicated webpages contextualizes the relation between tattoo and monument, leaving its appreciation and interpretation to the visitor. Importantly though, the parallelism that the pictures establish is not without implications. On one hand, the pairing of tattoo and monument implicitly associates both acts of remembrance, thereby aligning the personal narrative of the veteran—usually memorializing specific lost comrades—to the collective, national narrative of the memorial—instead premised on communal ideas such as sacrifice, patriotism, or heroism. On the other hand, the association of these two elements results in the temporal (re)inscription of the tattoo within the broader, linear narrative of the monument.

Working as a timing indexical in the individual narrative of the veteran, the tattoo supports a timing attempt that employs its owner's autobiography as the timing standard. The mediation of the war experience it enacts, then, keeps the individual storyline as its frame of reference, thereby attempting to order and synthesize it *within* the individual's autobiographical parameters. As a result, the conceptualization of wartime that emerges can be messy and open to negotiation—as suggested for Craig's tattoo, above—and still cohere with the veteran identity. Conversely, the rearticulation of the tattoo within the national narrative of the monument does away with the autobiographical specificity of the tattooed, adopting a different storyline—a national biography, for instance—as its timing standard. As a result, the chronic understanding of (war) experience is lost, replaced by a chronological rendering of wartime, the temporal parameters of which are no longer traced by the bearer of the tattoo but by a collective subject instead. War experience, in this new storyline, is 'a thing of the past', not of the unfinished present. Remembrance and memorialization—that is, past-oriented activities linked to a temporally bracketed wartime—become the focus of both monument and inked body, while the yet unfinished war experience is dispelled.

A comparable example is found in the 'Ink in the Lines' online archive by the Australian War Memorial (see *Ink in the Lines—Exhibition*, 2020). In this case, the physical exhibition is organized around a similar logic, with close-ups of military tattoos grouped into four macro-themes: 'identity and belonging', 'mateship and family', 'loss, grief, and commemoration', and 'healing' (ibid.). The online archive, instead, follows a different rationale. The website is not designed to 'tour' the virtual visitor around, instead providing a large set of different, unordered resources: a video of the physical exhibition with curators' comments, an interactive tool for visitors to upload the pictures of their military tattoos, twelve veteran-specific pages relating their tattoo narratives, edited footage of the veterans' video interviews organized by the same themes of the museum exhibition, and, last, some podcasts, video lectures, and articles (Australian War Memorial, 2020a). The articles and podcasts, in particular, foreground the existing link between war, militaries, and tattooing, historicizing body decoration practices and their relationship with war-making (see ibid.). Importantly, a similar socio-historical grounding of military tattoos is echoed in the exhibition. In the museum's physical display and its online video, as well as in some of the edited themed videos, veterans and their tattoos are portrayed next to pictures of older forms of body art, or tattoos from historically distant conflicts. Here again, then, as in the case of the Royal British Legion's archive, a purposeful alignment of individual and collective narratives is on display. Like the visual parallel with monuments explored above, through curatorship, the temporal parameters of war experiences are exchanged. In this case, curators have chosen to foreground the trajectory of military tattooing diachronically, tracing its 'evolution' as a memorializing practice in the Australian military (ibid.). Interestingly, the chronology within which tattooing is reinscribed is organized by conflicts: the First and Second World Wars, Malaya, Vietnam, First Gulf War, and East Timor (ibid., Memorial Articles, 'Inked: Tattooed Stories'). The result, however, is similar: chronic wartime is effaced and instead (re)narrated chronologically, in support of a distinctively nationalist narrative celebrating the Australian armed forces (see also Caso, 2020a).

## War as embodied experience and the problem of wartime

While the mobilization of military tattoos as a war object is a seldom studied phenomenon, embodied war experiences—including instances of their re-mediation in institutionalized war narratives (e.g., McSorley, 2012; Welland, 2017)—have been the focus of much scholarly attention in the last decade (see, for instance, Sylvester, 2011, 2012, 2013a, 2013b; McSorley, 2012, 2013, 2014, 2020; Parashar, 2013). Explicitly contesting the tendency to relegate war to a people-less, abstract domain made of states and supranational entities, these scholars sought, instead, to study war 'up from people' (Sylvester, 2013b: 484), foregrounding their 'physical, emotional, and social experiences' (Sylvester, 2013b: 2) with 'an explicit focus upon the *embodied* practices, structures of feeling and lived experiences through which war and militarism take place' (McSorley, 2013: 2, my emphasis). Situating their study of 'War Ink' precisely in this scholarly tradition, Dyvik and Welland (2018: 347) suggest that 'contemporary military tattooing should be considered as one such embodied and aesthetic expression of war experience', further specifying that '[m]ilitary tattoos are aesthetic and embodied sense-making practices that position the recipient not merely as subjected to wartime experience but as one who actively embodies and seeks to assert control over them' (Dyvik and Welland, 2018: 358). Conceived as a form of embodied timekeeping—that is a specifically *temporal* 'sense-making practice'—tattooing provides access not only to war experiences but also, as I argued above, to their distinctively embodied temporalities.

Building on William McNeill's work, Elizabeth Freeman (2019: 3) takes military drills to exemplify forms of 'temporalization of the body' in which 'flesh com[es] into meaningful embodiment and connectivity through adjusting itself to particular rhythms, that is, particular muscle memories whose accomplishment and automation fe[el] like a form of both selfhood and community'. Although not so specifically focused on their temporal facets, several studies of war experiences that highlight the corporeal dimension of war resonate with this position. Swati Parashar (2013: 619), for example, finds that '[t]he smell, taste, sounds, touch, pleasures and pains of war linger within and beyond the war', extending the claim of the endurance of certain, war-related, corporeal dispositions to war-touched bodies, beyond armed forces. Zeroing in on soldiering bodies, and specifically on infantrymen, John Hockey (2009) finds soldiers' sensorium and spatial practices to be moulded via training and combat, to the point of transformation. Focusing on a specific sense, Kevin McSorley (2020: 155), argues that 'it is not simply the marshalling of pre-existing and stable sensual worlds that occurs in wartime. Rather, sensory capacities become transformed and attuned to various aims and objectives', in a process that resonates with Wool's (2015: 131–155) idea above that soldiers are radically changed by war.

On account of their temporal protractedness, these 'transformations' arguably lie at the centre of (at least) some chronic understandings of wartime. Certainly, they testify to the importance of acknowledging (war)time pluralism and problematizing univocal re-mediations of wartime as finite. First, an 'embodied sociology of war' (McSorley, 2014) suggests that theorizing wartime as temporally bounded is analytically inaccurate. War experiences testify to the existence of war in peacetime and of peace in wartime (see Parashar, 2013: 618) and 'an analytic focus upon the body tends to render any clear demarcation of discrete war zones and times problematic, emphasizing instead the enactment and reproduction of war through affective dispositions, corporeal careers, embodied suffering and somatic memories that endure across time and space' (McSorley, 2013: 2). In short, once people and their attendant war experiences are foregrounded, war's spatio-temporal boundaries are shown to be porous, as war is found to exceed a conflict's conventional beginning and ending.[5] Unsurprisingly, the idea of victory is not found frequently in these studies. Keeping chronic and chronological embodied temporalities of war distinct, however, further suggests that there are at least two ways in which war experiences challenge the notion of a temporally bracketed wartime. By predating and outliving a given conflict's conventional start and end dates, some war experiences prove abstract war chronologies inaccurate (see, for instance, Niromi, Babu, and Ranjini's experiences in Parashar, 2013). Other experiences, however, challenge the notion of war's temporal finitude altogether.

Indeed, and second, several scholars have underscored that wartime's temporal finitude proves unfaithful to certain war experiences. Kenneth MacLeish (2013: 7) for instance, laments that when it comes to lived experiences of soldiering, 'the language available to help us understand what is happening ... is rather impoverished'. Sarah Bulmer and David Jackson (2016: 28) articulate this discomfort more succinctly, arguing that the trouble with extant accounts of war experiences is a problematic reliance on 'the idea of beginnings and endings', in itself inadequate to grasp 'experience[s that] disrup[t] linear understandings of time, place and identity' (ibid.). Echoes of a similar position are found in the personal accounts on display in the exhibitions of military tattoos analysed above. For some of these veterans, as I have shown, the war experience is never over, as war lives on *in* and *through* them and their bodies, beyond training, combat, and deployments. War, in these cases, unfolds in a chronic temporality.

Nevertheless, publicly sanctioned war narratives that are unfaithful to specific lived experiences, or even analytically inaccurate, are not new. In the aforementioned ethnography of Walter Reed Army Medical Center, Zoë

[5] As a result, war experiences are retrieved from, inter alia, museums (McSorley, 2012; Welland, 2017; Sylvester, 2019), memorials (Sylvester, 2018), cemeteries and mass graves (Kwon, 2008; Sylvester, 2019), military communities and recovery facilities (MacLeish, 2013; Wool, 2015), but also conflict textiles (Andrä et al., 2020), memoirs and war fiction (Sylvester, 2013b, 2019; Dyvik, 2016; Welland, 2018), fashion objects (Bide, 2021), sociotechnical apparatuses, including weaponry (Brandimarte, 2023), and videogames (Jarvis and Robinson, 2021).

Wool (2015) reflects extensively on the representation of soldiers' injured bodies. Discussing the different public events taking place at Walter Reed, Wool (2015: 82–95) focuses specifically on the intermingling of personal and collective accounts of injuries and wounds that these events engender: 'in these displays Walter Reed becomes a place where the moral economy of war can jump scales, from individual bodies to the nation at large, from the meaning and value of wars of nations to the qualities of an individual solder's body and life' (Wool, 2015: 83). As she notices, though, the mobilization of bodies in collective narratives is not without implications. As part of the process, not only are veterans' bodies 'recruited for more passionate expression of national wartime feeling' (ibid.), but their personal narratives—and related war experiences—simultaneously effaced. Relating one of such events—namely, a public visit of President Lyndon B. Johnson to the facility in May 1966—Wool (2015: 84) hints boldly at this dynamic:

> [These public displays] are never precisely about twenty-five-year-old Anthony J. Troilo. They are never precisely about those soldiers whose bodies leave President Johnson breathless. Walter Reed as a national stage iconizes these particular soldiers: encounters with them are exemplary and allegorical, unfolding in a register of citizen, soldier, and nation. They are embodied, but not exactly personal.

Unsurprisingly, patients at Walter Reed are not always eager to participate, often opposing explicitly the political folding of their injured bodies within narratives of heroism and national pride. On a similar note, Julia Welland (2017: 524) finds displays of individual war experiences at the Imperial War Museum to be 'crucial both for how wartime violences [are] mediated and understood, and for putting in place the conditions of possibility for their continuation'.

Re-mediations such as those instantiated by the exhibitions of military tattoos under analysis here are no less problematic. By reinscribing war experience within overarching narratives of war as a temporally discrete event, these exhibitions end up effacing both the distinctively chronic temporality of certain embodied experiences of war, and their associated understandings of wartime. As a result, wartime is subtly, but actively, reified as unitary, temporary and temporally bracketed. As discussed in the Introduction and Chapter 1, far from being an innocuous abstraction facilitating the representation of war, such a monolithic notion of wartime is politically problematic and violent. As feminist scholars noted time and again, for example, 'the idea of war as a discrete event with an identifiable beginning and end and a clear location does not stand up to scrutiny when we examine women's experience before, during, and after war or when we think about what is required to wage war' (Sjoberg, 2013: 274; see also Gray, 2019: 194–195; Kirby, 2020: 223).

Chronological renderings of wartime risk obscuring how 'war is present beyond the war front and *beyond wartime*, in and among apparently pacific social, cultural, and economic relations' (Barkawi and Brighton, 2011: 132, my emphasis; see also Barkawi, 2011: 707; 2016: 201–204; Basham, 2013; Sjoberg, 2013: 271–276; Gray, 2019: 195). In fact, these and other similar exhibitions are one way in which wartime can be upheld as a state-led discourse that sanctions exceptionalism and the bending of the rule of law (Dudziak, 2012). Victory tropes, in this context, have similar effects: they contribute to the circulation of an understanding of wartime that 'has temporal boundaries on both sides, so that entering a "wartime" is necessarily entering a temporary condition' (Dudziak, 2012: 15; see also Hom and Campbell, 2022a; McIntosh, 2022a).

Several of the war experiences on display in these exhibitions of military tattoos testify to the inadequacy of a clearly demarcated wartime notion. And yet, exhibitions place these wars in spatio-temporally distant and defined locations or insist on the difficulties of transitioning back to 'civilian life' (see Dyvik and Welland, 2018: 353), subtly, but actively, reinforcing a problematic war/peace(time) binary. Firstly, it follows, deployments of victory that assume its capacity to demarcate wartime from peacetime, even when the point is victoriousness, are just as complicit to these mechanisms. As a result, and secondly, to enquire into victory's temporal politics becomes both a matter of analytical and political urgency. Just like these exhibitions, victory utterances that are inattentive to the concept's temporal assumptions—whether wilfully or inadvertently—risk being unfaithful to, or even effacing, certain lived experiences. On a broader level, they also normalize wartime's temporariness and exceptionality.

Lastly, war's temporal pluralism—or heterotemporality (see Hutchings, 2008)—needs to be the baseline of victory research. How can victory inform our understanding of war but not the experience of those who fight it? The assumption of wartime's temporal finitude, indeed, is often constitutive of victory, defeat, and other concepts associated with war endings—that is, for military victory to mark the end of war, the temporal boundedness of war needs to be either presumed or carefully crafted. These exhibitions are also a powerful reminder that in the study of victory, references to the end of war, finality, and decisiveness should be carefully unpacked. Importantly, I do not wish to uphold military tattoos and related narrative as the *only*, or even the main, source for the study of the war experience. Nor do I wish to maintain that victory is not a relevant category for the study of veterans' encounters with war, or that temporal unboundedness is a necessary feature of all individual war narratives. Rather, my contention is that taking stock of the tension between wartimes highlighted above is empirically and analytically critical. Indeed, it signals that supposedly victorious bodies are not necessarily so conceived by the selves inhabiting them—at least when victory is endowed with the temporal capacity to bring conflicts to an end. Centring time in the study of soldiers' embodied experience, in other words, casts hegemonic understandings of victory into doubt.

## Conclusion

Grappling with the temporal complexity of war experiences, in one of their staged conversations, scholar Sarah Bulmer and war veteran David Jackson (2016: 28) explain the problem as follows:

> DAVID: … To get a sense of who I am, and my experiences of living with war, you have to understand my journey.
>
> SARAH: Where does that journey begin?
>
> DAVID: There is a problem with the idea of beginnings and endings. My journey isn't linear. … My sense of war, even where the particular events are in the temporal past, is never distant. Experience disrupts linear understandings of time, place and identity. You walk in with some sense of me based on your prior knowledge of me. … However, a sense of war can't be packaged neatly for others' consumption or understanding. So, fundamentally, the challenge is raised in the question I ask you: Can you, as a researcher, adequately comprehend my sense of war?

This chapter has disputed the idea that military victory brings war to an end in the sense that it admitted the possibility that wars might not end in the first place. Thinking alongside Bulmer and Jackson, the chapter has also offered a conceptual tool to grasp war experiences, such as David's journey, beyond 'the idea of beginnings and endings'. Sitting comfortably with 'experience[s that] disrup[t] linear understandings of time, place and identity', this theoretical innovation—a conceptual distinction between chronic and chronological timing outcomes—enables scholars to think about wartime in its plurality and to apprehend the multiple temporalities of war.[6] Indeed,

[6] I wish to acknowledge here that theoretical reflections such as those offered in this chapter still constitute forms of narrational re-mediations of war experience—or even timing attempts—that mobilize military tattoos and the bodies of their bearers. I do not claim to have offered a more faithful

not only has this chapter offered empirical evidence of wartime's temporal pluralism—or heterotemporal nature (see Hutchings, 2008)—it has also shed light on some of the formal mechanism at the heart of the notion's homogenization and normalization.

Taking this intuition as its starting point, the following chapter wrestles for a bit longer with the idea of victoriousness and the temporal discrepancies it gives rise to. While a chronic wartime and the inadequacy of victory might be representative of the experience of the veterans featured in these exhibitions, victory remains an enticing promise for recruits across militaries (Daughtry, 2015: 133), and some individuals—including soldiers—might well talk about it. Through the analysis of victory monuments, the next chapter analyses collective identity narratives and shows how victoriousness can even be passed on from collectivities to individuals—temporal inconsistencies between these subjects' storylines notwithstanding.

representation of any of the war experiences on display in these exhibitions nor that by adopting the chronic/chronological distinction researchers will fully 'comprehend [veterans'] sense of war'. Rather, by advancing the study of the temporalities of war experience, this chapter sought to offer a tool to stay with the murkiness and temporal unboundedness of certain experiences, without 'packag[ing them] neatly'.

# 5

# Monumental Victories

## Myths, Collective Identities, and Social Orientation

### Introduction

This chapter rounds up the examination of victory's temporal politics through the analysis of one last, widespread victory practice: the building of monuments. I scale up my analysis of the link between victory and time once more, attending to three victory monuments around the world, beyond the Anglosphere: Baghdad's Hands of Victory, the Bronze Soldier of Tallinn, and the Victory Gate in Bolzano (Italy). These monuments were built at different points in the twentieth century to celebrate or memorialize specific military victories—in the Iran–Iraq War, Second World War, and First World War respectively. They display important differences in their architectural design, how they symbolize or embody victory, and the 'level of contestation' they have attracted since their construction. Nevertheless, in one way or another, they all reproduce the hegemonic understanding of victory as a material outcome of armed conflicts challenged throughout this book. Victory monuments, in other words, further demonstrate that, contrary to established wisdom, victories are made inside and outside the wartime battlefield.

By looking closely at these monuments, the chapter uncovers three additional insights about military victory and its relationship with time. First, victory is ubiquitous in public and specialist discourse because it works as a provider of social orientation: it can *time* war *and* identity narratives, at both the individual and collective levels. Much of the contestation surrounding victory monuments, indeed, is essentially identitarian: it coalesces around the victories they signify precisely because these have implications for individual and collective identities. Second, victories can be produced, given material substance, and be effectively popularized among publics no matter how fictional, mythical, or 'staged' they are when assessed against military developments on the ground. In fact, monuments are often employed to quite literally cement specific victory narratives, constituting material proof that a

*Military Victory Beyond the Battlefield*. Mirko Palestrino, Oxford University Press. © Mirko Palestrino (2025).
DOI: 10.1093/9780198955818.003.0006

given victory has happened—especially in the absence of 'facts'. In turn, identity narratives mobilize these same monuments, sometimes to buy into that very victory narrative and stabilize a subject's biography, but sometimes to try to distance themselves from that victory, or even challenge it altogether. Third, and relatedly, a plethora of narratives, therefore, partake in the social construction of military victory and its temporal qualities. Whether it is an outright narration of a given conflict, a counter-narrative that seeks to challenge the historical consensus around a war, or even an identity narrative that merely *times* victory against the biographical coordinates of one or more subjects, they all contribute to the production of victory and its temporalities.

I start with a brief, theoretical discussion of the relationship between victory, monuments, and identity. I explain how identity politics provide fertile terrain for the making, contestation, and reimagination of both monumental architecture and victory's meaning and temporalities. The chapter then moves to three empirical examples. The analysis of Baghdad's Hands of Victory shows how a monument can be used to materialize a victory amid an ongoing war and in the absence of favourable military developments. I also demonstrate that by building a 'didactic monument', Saddam Hussein contextually sought to popularize a specific version of history and elicit identifications with it. In turn, the case of the Bronze Soldier of Tallinn illustrates that popularization and identification can happen also in the case (and by means) of contestation. In fact, the Estonian 'war of monuments' offers a clear example of how identity narratives not only tend to converge around a victory monument, but also reflect back on the very victory that monument symbolizes, shaping public perception about it and even casting doubt on the established historical record. Finally, I turn to the Victory Gate in Bolzano. This is an example of contestation going as far as to lead to state-sanctioned, counter-monumental practices—or an official attempt to amend the historical record. Beyond further substantiating the points above, this example uncovers the resilience of hegemonic understandings of victory as a material, time-altering event: even in the wake of structural interventions that seek to change the meaning of a victory monument, the temporal qualities associated with victory can remain unaltered.

## Biographical victories: monuments as orientational devices

The making of monumental art is arguably one of the oldest and most widespread victory practices. In ancient Greece, for example, victory in war was customarily marked by the erection of a battlefield trophy: an artefact usually composed by either a pole or a tree trunk to which the enemy's arms were dangled or fastened (Stroszeck, 2004: 303–306; O'Driscoll, 2020a: 49; see also Nováková and Šályová, 2019). Trophies not only served a memorializing function but were integral to the making of victory: the erection of a trophy was a necessary condition to mark the end of war. Indeed, as Cian O'Driscoll (2020a: 51) explains, failure to produce a trophy 'was a cause of great consternation and led to further wrangling' (ibid.). Similarly, the world famous ceremonial parades known as Roman triumphs normally culminated at a triumphal arch built ad hoc to celebrate a general and memorialize his victory (see McCormick, 1986; see also Beard, 2007; and O'Driscoll, 2020a: 45–49).

By no means an exclusive feature of ancient Greece and Rome, victory monuments have been—and still are—popular across histories and geographies (see, for example, Leisten, 1996 on Islamic medieval societies, or Brüggemann and Kasekamp, 2009 on Soviet republics) for at least two intuitive reasons. First and foremost, monuments are public displays of power, 'they "re-present" political entities, states, and statesmen in a very literal sense: making them "present" in public spaces' (Hölscher 2006: 27). Second, monuments are also pivotal to the spatial production of social identities and imaginaries—or, as Karen Till (2003: 290) eloquently put it, monuments 'attempt to "fix" time and identity through the material and symbolic qualities of place' (see also Nora, 1989: 13). When it comes to military victory, then, monuments not only celebrate and memorialize, but also offer tangible proof that victory has been achieved: they *materialize* that victory, just like Greek trophies used to. Notably, they contextually open up space for individuals and collectivities to anchor their identity narratives to them, for example, by including whatever episode or value a given monument symbolizes in their biographies, or even by opposing and contesting it.[1]

For these very reasons, however, monuments also tend to stir up crowds, especially in the wake of large-scale violence—like the wars that precede the construction of victory artefacts. According to Till (2003: 297, my emphasis), indeed, places of memory 'punctuate and create symbolic space, and function as *nodes of collective politics* at and through which notions of identity (such as race, class, gender, and the nation) are performed and contested'. Unsurprisingly, scholars have turned to monuments and the vicissitudes around them

[1] As discussed in Chapter 3, for instance, the Cenotaph in Whitehall did not only work as 'proof' that the First World War was finally over, but it also became central to narratives of national mourning and, ultimately, patriotism and Britishness.

time and again for the study of identity (see, for instance, Osborne, 1998; Forest and Johnson, 2002; Mitchell, 2003). Pierre Nora (1989: 19), for example, famously conceptualized monuments as *lieux de mémoire* to emphasize the key role they play in identity politics on account of their 'capacity for metamorphosis, an endless recycling of their meaning and unpredictable proliferation of their ramifications'. Forest and Johnson (2002: 525), even concluded that 'the analysis of *lieux de mémoire* (places of memory) provides an ideal way to trace underlying continuities and discontinuities in national identity politics'.

While I am not primarily interested in identity politics, these insights on the identitarian nature of monumental art are particularly relevant for the study of victory and its temporal qualities. Indeed, for Vale (1999: 391–392) the 'media dimensions of all politically affiliated architecture' is constitutive of monuments and their urban context, to the point that we better conceive of them—and therefore approach their study—as 'visually enhanced narratives'. And just like victory practices, identity narratives can, and have been, theorized and understood as *timing* operations (see, for instance, Hom and Solomon, 2016). Attending to the abundance of identity narratives that concentrates around victory monuments, then, it is possible to trace, discern, and empirically analyse the various temporal ordering processes that underlie these narratives, as well as their effects on victory and its temporal parameters. Broadly speaking, these effects are of two kinds. First, the joint temporal work of victory and identity can explain the ubiquity of victory in public discourse—and especially the fact that victory is often mobilized as a key nodal point of national biographies and other collective identity narratives (see Hall, 2022; Steele, in Hartnett et al., 2022). Second, biographies and other identity narratives do not only draw on pre-existing understandings of victory, but also affect, and sometimes even alter, the temporal parameters of the concept or specific victories.

In his work on international orders and identity, Felix Berenskoetter (2014: 264) suggested understanding 'the world as composed of multiple overlapping biographical narratives'. '[A] biographical narrative', he specifies, 'is not a record of everything that ever happened to the Self, but highlights experiences that matter' (Berenskoetter, 2014: 269). In particular, biographies fulfil an orientational purpose, that is, they 'provide the Self with knowledge about its place in "the world"', and help 'situate the Self and delineate its existence in time and space' (ibid.). I first want to suggest that military victories are exceptionally well suited to fulfil this orientational purpose and that it is for this reason that references to victory are frequent in identity narratives.

There are at least two senses in which victory provides social orientation—or, offers a subject 'knowledge about its place in "the world"'. Firstly, on account of its relational character—or its intuitive link with the idea of defeat—victory captures the relative positioning of the victor(s) vis-à-vis the vanquished. Here, orientation takes place in oppositional terms, resembling traditional accounts of identity constitution that emphasize relationality in general, and othering processes in particular (see, for instance, Neumann, 1996). Embedded in the concept, then, the situational character of victory lies in its capacity to be simultaneously about somebody/something *and* somebody/something else. Secondly, binary juxtapositions such as victor/vanquished might prove useful biographically beyond their immediate antithetical structuring. Indeed, their mobilization in narratives can help corroborate or challenge meaning-making efforts of one's sociopolitical situatedness more broadly.[2] A conspicuous example here is the antithetical positioning of the allied powers of the Second World War to the evils of Nazism and tyranny. As part of this hegemonic discourse, victory simultaneously entails not only the defeat of Nazi Germany but also the elevation of the victors to a position of moral superiority. In turn, this ascendancy is biographically associated with international 'responsibilities' and privileges, as epitomized, for instance, by the political economy of the United Nations system.

The temporal parallel to these ideas is that victory offers both *internal* and *external* temporal stability to a subject. The former can be understood as a 'sense of coherent self' (Hom and Solomon, 2016: 27), or the orderly unfolding of the different continua of change that are normally coordinated and synthesized in an identity narrative (see Hom, 2018a, 2020; Palestrino, 2022a). To stay with the example of the Second World War above, Germany can be reborn after its defeat and the Nazi regime confined to a past from which to actively seek distance. As a temporal device, indeed, victory allows for different (re)orientations within the storyline of the Self, such as new beginnings and endings. Notably, victory is also constitutive of the broader

[2] Importantly, biographical narratives do not produce victories in a vacuum. When employing victory as a sense-making device they do so against the backdrop of pre-existing understandings of the notion—i.e., victory symbols can time too (see Chapter 1). As a result, for instance, a biographical narrative does not need to stress or explain that a given victory brought a conflict to an end. Instead, it can simply assume so, and thereby silently reproduce the idea of temporal finality associated with victory that was problematized in previous chapters. Moreover, when unpacking the role that victory plays in the construction of identity narratives, especially in terms of its temporal politics, the existence of overlapping timing operations needs to be kept in mind. While it is heuristically convenient to think about these narratives as separate, their mutually reinforcing character ought to be concomitantly emphasized. This is especially evident when foregrounding victory as the object of inquiry. Biographical narratives, other victory utterances, as well as specialist studies of military victory all contribute towards informing our understanding of the notion. In turn, its deployment in different narrational enterprises carries with it this complex web of information.

spatio-temporal 'environment' within which it takes place and, therefore, concomitantly brings *external* temporal stability to identity plots. In particular, victory does not only concern the victor and the vanquished, but also the social time-spaces they share. Indeed, as any other temporal operation, an identity narrative is a device through which a social actor is able to synthesize and coordinate *external* events to make sure that 'the time we inhabit is consistent, reliable, and orderly rather than some onrushing and untamable river of unintelligible events' (Hom and Solomon, 2016: 31).

Following the theoretical set up from Chapter 1, however, this temporal flow should not be conceived as a *thing*, a given to be accounted for and with which a subject needs to coordinate. Rather, it is a process upon which that same subject has some leverage, or social agency. When timing through an identity narrative, not only can social actors 'adjust' their storylines for the sake of inhabiting a stable—or at least organized—(*external*) temporal flow (Hom and Solomon, 2016; see also Solomon, 2014) but they can also act on it, trying to influence the ways in which it unfolds. For example, by buying into or contesting specific victory narratives, subjects can, for example, uphold some victories—such as the victories in the World Wars—as important turning points in the histories of regional or world political affairs. In turn, these accounts go on to inform broader historical chronologies, for instance by reinforcing or disputing their credibility, or by drawing continuities between a given conflict and another.

These dynamics have very tangible implications. Not only do they explain the recurrence of victory in identity narratives, but they also shed some valuable light on the social construction of victory and its temporal parameters. Indeed, it becomes apparent that narratives that are not primarily about victory can still inform what victory means, as well as the link between victory and time. In fact, information about victory and its relationship with time is found in, and therefore supplied by, both identity narratives that deploy victory as an orientational device *and* narratives that explore or recount victory directly[3]—while not being concerned with articulating the storyline of a given social subject.[4] Put differently, victory can *time* both when used as a narrative *and* as a symbol.

[3] Examples of identity narratives include biographies of specific individuals, all the way up to national and supranational collectivities, including smaller groupings such as political parties or transnational movements. Historical accounts of specific victories, studies of the concept like those engaged in this book, or even foreign policies aimed at victory such as Obama's and Bush's discussed in Chapter 1, in turn, are all examples of 'direct' recounts of victory.

[4] This is a heuristic differentiation between a set of complex narratives that parallels Hom and Steele's (2020) distinction between biographies and cosmographies. Reading the social environment inhabited by different collectivities—e.g., 'the world', or 'the international'—as a (narrational) collective identity project in itself, these authors reflect on its relationship with the different collective identity narratives

These ideas are particularly evident in the case of victory monuments. On the one hand, victory monuments are, by definition, victory symbols—i.e., they embody and signify victory. Therefore, monuments offer a *direct* account of victory. On the other, victory monuments are, like any other monument, loci of identity politics. The identity narratives that tend to cluster around them are also, and often *indirectly*, about victory. As an empirical site, then, victory monuments have the great advantage of capturing these complex interactions between overlapping and competing timing operations. Just as the same tattoo(s) can facilitate the articulation of multiple war stories, so too can a victory symbol like a monument sit at the centre of overlapping narratives and attendant timing attempts. In short, victory monuments also work as timing indexicals. I now turn to three such empirical sites, drawing out different, and complementary insights on victory from each of them.

## Baghdad's Hands of Victory: mythical victories and identification

First announced to the public on 22 April, 1985 and officially inaugurated on 8 August, 1989 (Makiya, 2004: 1), the 'Victory Arch' in Baghdad is Saddam Hussein's most renowned victory monument, and perhaps one of the most widely known of the genre, all over the world. The monument consists of two, perfectly equal, gigantic victory arches that face each other at a distance of 650 metres and delimit both entrances to the Parade Ground—or Grand Festivities Square—in central Baghdad (Van de Ven, 2020: 118–119). While arguably drawing on the classic architectural trope of the victory arch from ancient Rome (see Van de Ven, 2020), the monument is visibly different in style and design. For example, rather than replicating the traditional geometrical design of Western triumphal arches, each of the titanic structures features two massive, crossed swords held by Saddam's forearms,[5] emerging from the ground and forming an archway forty metres tall at its apex. On the

unfolding within it (ibid.). In particular, Hom and Steele (2020: 327) find collective social actors' biographical narratives to be 'the building blocks of a general effort to configure their wider environment as a coherent, whole, world'. However, they also find this 'environment' to be parallelly informed by other non-actor-specific biographies, that is by narratives of the 'international realm per se', which they term 'cosmographies' (Hom and Steele, 2020: 329). Rather than constructing the social, the world, or the international only indirectly—i.e., as the background of specific actors' biographies—these accounts are aimed precisely at providing an outright narration of it. While certainly a simplification, I find the distinction helpful insofar as it captures how a notion—like 'the international' or 'victory'—can be simultaneously crafted via direct *and* indirect engagements with it.

[5] As Makiya (2004: 8) famously put it, 'these are not anybody's arms, but the President's own'—that is, in the making of the structure, a plaster cast of Hussein's arms was used.

invitation card produced for the monument's opening ceremony, the Victory Arch is described as composed of five main elements:

> (1) The Exploding Ground: made from reinforced concrete with enemy helmets scattered above; (2) The Forearm and the Grip: cast in bronze and weighing 20 tons each, fixed on to a steel frame also weighing 20 tons; (3) The Sword: the sword of Qadisiyya was used with a slight bend in order to give the arched shape. It is made of stainless steel. Each sword weighs 24 tons. The grip and the martyrs' dome [?] were cast in bronze and weigh 4 tons each; (4) The Net: cast in bronze, each net contains 2,500 helmets taken from the Iranian enemy; (5) The Flag: made from stainless steel and rising 7 metres above the meeting point of the two swords. The stainless steel used in the project (the swords) was made from the weapons of the martyrs and weighs 42.6 tons.
>
> **(Executive Committee for the Victory Arch, cited and translated in Makiya, 2004: 2)**

Unsurprisingly, then, the monument is also known as the 'Swords of Qadisiyya', the 'Hands of Victory', and the 'Crossed Swords Complex' (Van de Ven, 2020: 118–119).[6] However, it is the original name, 'Victory Arch', that most explicitly encapsulates the monument's intended purpose and politics.

Erected to celebrate 'the great victory' (Executive Committee for the Victory Arch, cited and translated in Makiya, 2004: 2) in the Iran–Iraq War 1980–1988, the monument, I suggest, was explicitly designed to (a) support the production of a military victory in the absence of an evident military success on the ground, and (b) inscribe that victory within the national biographical narrative of Ba'athist Iraq. Together, these two temporal operations sought to foster nationalism by eliciting individual identifications *with* Iraq because of that victory—i.e., rallying people around the flag and convincing them that victory had been achieved.

On the one hand, this is unsurprising. Indeed, group identification practices are key to nationalism. According to William Bloom (1990: 54–75)—arguably one of the first scholars to translate and apply identification theory from social psychology to international politics—'nation-building *requires* that the mass of individuals make an identification with the nation-state' (Bloom, 1990: 61, my emphasis). Identification, in turn, is elicited through symbols, ideas, and rituals that 'present a mode of behaviour or a set of attitudes the adoption of which will enhance identity and psychological security'

[6] For a brief discussion of the issue of the monument's name see Van de Ven (2020: 118–119), especially footnotes 24 and 25.

(ibid.) without an individual being required to experience what a specific ritual or symbol is about. Social rites around national wars, for example, serve precisely this purpose: fostering individual identifications with the nation, including among those who did not fight (Bloom 1990: 64–67).

On the other hand, the use of a victory monument to achieve these ends is particularly salient. By treating Baghdad's Hands of Victory as a timing indexical, it becomes clear that temporal operations can be sustained and apprehended through the analysis of design choices and celebratory practices associated with the monument.[7] In particular, I argue that this monument worked as a victory symbol with a double function: timing victory in the Iran–Iraq War amid ongoing hostilities, while simultaneously timing Iraq's biographical narrative *through* that victory.

The intention to, first and foremost, erect a monument that would speak to (a specific version of) Iraqis' national identity, rather than merely celebrating a historic event, was manifest in Hussein's speech of 22 April, 1985—an excerpt of which was reproduced on the 1989 opening ceremony invitation card already cited above. The quote is worth reporting in full:

> The worst condition is for a person to pass under a sword which is not his own or to be forced down a road which is not willed by him. From this, and because brave Iraqis have recorded the most legendary exploits in defence of their land and holy beliefs against the Persian enemy in the era of the liar Khomeini . . . we have chosen that Iraqis will pass under their fluttering flag protected by their swords which have cut through the necks of the aggressors. And so we have willed it an arch to victory and a symbol to this Qadisiyya.
>
> **(Hussein, cited and translated in Makiya, 2004: 2)**

Not only is the monument explicitly *for* Iraqis, who 'will pass under their fluttering flag',[8] but the circumstantial evidence about the event—or war outcome—that the Victory Arch is meant to celebrate suggests that the event itself, the victory, is far from clear and self-explanatory. First, the original speech was delivered in the midst of the Iran–Iraq conflict in 1985, when the clash of arms was far from being over. Nevertheless, in incendiary terms, Hussein refers to the enemy as a liar and an aggressor, whose fate—that is, an utter defeat at the hand of the Iraqis—has already been written. As Makiya (2004: 10) has suggested the 'conception [of the monument] therefore precedes the

[7] An analysis of material victory symbols such as monuments, in other words, also shows the benefits of studying time extratextually.

[8] To be noted here is also the intention to build a monument explicitly designed for military parades, to which the quote hints. This is also evident in the spatial positioning of the arches at both entrances of a parade ground.

reality it is meant to commemorate'. In turn, according to Van de Ven (2020: 112), this was a strategic political move to gather support in the wake of a legitimacy crisis exacerbated by the conflict itself. In other words, the monument and accompanying practices seek to explicitly time the war—and do so via victory.

Second, even in 1989, when the opening of the monument to the public was supposed to commemorate 'the occasion of one year passing since the day of the great victory' (Executive Committee for the Victory Arch, cited and translated in Makiya, 2004: 2), the classification of the ceasefire as a 'great victory' for Iraq was—and still is—largely disputed, at least in scholarship. Historian Efraim Karsh (2002: 85), for instance, described the war outcome as 'inconclusive' and 'with no clear victor'—a position which is reiterated by both Van de Ven (2020: 130) and Makiya (2004: 15–16). In a similar vein, Nadia El-Sayed El-Shazly (2016: 337) noted that 'while neither had triumphed ... Baghdad and Tehran both declared victory'. Nonetheless, a majestic monument to the supposed victory was erected and publicly celebrated. According to Kanan Makiya (2004: 15), the monument met 'a deeply felt collective need to believe in a mythology such as victory in the Iraq-Iran war'. Notably, that a collective identity narrative is mythical in nature does not hinder individual identifications with that narrative; if anything, vagueness is constitutive of collective visions, allowing for the enlistment of a larger number of individuals. According to Felix Berenskoetter (2014: 271), indeed, a direct link between an individual experience and a collective myth is not needed for identification to occur. 'Humans', he writes, 'seek to inscribe themselves into larger, lasting spatio-temporal structures in an attempt to give meaning to their contingent existence' (Berenskoetter, 2014: 269)—and, in fact, to that end, collective visions are often centred around mnemonical and mythical past experiences that 'must be sufficiently vague ... to allow a large number of individuals to adopt it' (Berenskoetter, 2014: 273).

Besides constructing victory as an event (or outcome) *in medias res*, the orientational function of the Victory Arch is also carried out via othering practices surrounding—that is, embedded in as well as reinforced and reified by—the monument and the victory it seeks to represent. Here, I am not only referring to the explicit othering *in* and *of* Hussein's discourse, evident in both the 1985 speech and the rest of the 1989 invitation card (see above and Makiya, 2004: 1–4). Instead, I want to suggest that othering practices feature *in* the monument, with elements of its design attempting to materialize these logics and consign them to eternity (on this point, see, Edkins, 2003: 83; Lisle, 2016: 128–129; Neumann 2018). The incorporation of Iranian helmets within the monumental structure is perhaps the starkest example of this operation. Two main details stand out in this regard. First, the very nature of the 5,000 Iranian helmets, that is, 'war debris' (Makiya, 2004: 3), or original pieces of the enemy's defensive gear sacked from the battlefield throughout the eight years of conflict, confer authenticity to the monument and the victory it depicts.

Second, their display in large, bronze nets at the base of the structure, is an evident attempt to qualify that 'debris' as proper spoils of war, as booty. Here, the inscription of victory within the monument is realized through a display of the phantomatic body of the enemy and the ostensive portrayal of defeat on account of its close relationship with victory. Importantly, though, this centres the enemy *within* the victor's narration, thereby solidifying that process of othering according to which the vanquished is attributed an inferior social standing. Similarly, the geographical positioning of the arches, namely their location on both ends of the Parade Ground, is meant to reinforce this logic, perpetually exposing the spoils—dominated by its owner, an exceptionally vigorous and powerful Iraq—to every onlooker. Moreover, the monument's primary architect, Khalid al-Rahal, emphasized how both arches are oriented so that the swords point to 'the two great enemies of Arab Islam' (Van de Ven, 2020: 117–118), namely Israel and Iran, thereby reinforcing the intended conflictual politics of the arches. Once again, the monument is not only preoccupied with commemorating or celebrating a victory. That victory, instead, is simultaneously employed to provide a collectivity with a 'body', position it socially, and offer that orientation up to those individuals who identify with it.

Another example of this active attempt to enshrine identity logics of Self and Other *in* and *through* the monument, then, is precisely the rich symbolism of the swords. Indeed, as well as being positioned strategically, the swords are significant for at least two additional reasons. First, and as pointed out by the monument's description on the invitation card (see quote above), the stainless steel used for the blades 'was made from the weapons of the martyrs' (Executive Committee for the Victory Arch, cited and translated in Makiya, 2004: 2). As Kanan Makiya (2004: 8) explained, according to the Iraqi official discourse, the steel was obtained from melted Iraqi weaponry, mainly machine guns and tanks, employed during the conflict. Thus, the same claim to authenticity introduced above with reference to the Iranian helmets can be extended here, with specific reference to the medium of the blades. Second, the official description of the arch also emphasizes how the swords reproduced in the monumental structure are not a generic type of sword, but 'the sword of Qadisiyya' (Executive Committee for the Victory Arch, cited and translated in Makiya, 2004: 2). In particular, the swords are meant to be those of Sa'ad ibn-abi-Waqas, the military commander of the Arab-Muslim army who, in AD 637, defeated the Persian Sassanian Empire in the battle of Qadisiyya (Makiya, 2004: 10–11). The idea, then, is to extend the othering process throughout a longer historical trajectory, identifying the Iraqi victorious Self with that of the Arab Muslim army, and the Iranian vanquished Other with the ancient Persian Empire. Here, it is worth pausing to consider, alongside this trajectory, the active attempt to inscribe the supposed victory of 1988 in a meticulously crafted biographical narrative of Ba'athist Iraq. In short, victory's timing function is forcibly on display once again: a sense-making device orienting social subjects and establishing chronological links between distant space-times.

First, and as can be noted from the 1985 speech excerpt above, Hussein employed the rhetoric of the Qadisiyya throughout the whole conflict (El-Shazly, 2016: 49; see also Makiya, 2004: 10–11). In fact, Ba'athist propaganda repeatedly referred to the 1980–1988 war as *Qadisiyyat Saddam* in an attempt to, on the one hand, cast an aura of holiness around the figure of Saddam Hussein, and, on the other, inscribe the original battle of Qadisiyya within the same historical trajectory of the Iran–Iraq War. On this account, not only is the Self/Other distinction given historical depth, as noted above, but most importantly, Ba'athist Iraq is attributed a rich historical—although mythical (see Makiya, 2004)—origin, with direct links being traced to the ancient Arab Muslims. Second, and relatedly, ceremonial practices related to the monument worked to further corroborate the narrative and lend robustness to the attendant timing attempts. In particular, at the opening ceremony[9] in 1989, Saddam Hussein 'was televised riding under [the two arches] ... on a white stallion' (Makiya, 2004: 4). As Makiya (2004: 11–18) explained, the choice of the horse revolves around its crucial symbolism, through which Saddam Hussein, a Sunni, is linked to Husain the martyr, son of 'Ali. In turn, the choice was supposed to corroborate Hussein's (the president) claim—made in a public speech in 1979 and reiterated through street art and in Hussein's biography—to be a direct descendant of 'Ali (ibid.).[10] The idea, here,

[9] Interestingly enough, in her analysis of Saddam's Victory Arch and ancient Rome triumphs, Van de Ven (2020: 119) has compared the opening ceremony to North's idea of a 'collective emotional event', Turner's 'social drama', and the Romans' '*evidentia*', that is 'a tactic that induces "emotional sharing"'. As she put it, the ceremony was meant to 'collectively perfor[m] a specific identity' and thus bring that collectivity into being.

[10] In the political economy of the ceremony, though, the choice to reiterate this idea is yet another instance of the specific oppositional character of the victory narrative, expressed in identitarian terms. As Makiya (2004: 17) has noted, Saddam Hussein 'drives through his arches not on a modern tank, but on a white horse even though he is intent on overthrowing a Shi'ite revolution. On another level, therefore, his monument, like his biography, is a bold political assertion in the teeth of known facts (like his Sunniism, his nationalism and the fact that the war's outcome was a standoff, not a victory)'.

as Makiya (2004: 11) has put it, is that 'national unity prevailed in the face of foreign aggression' as 'this is the Ba'athist line on the war'. Through the figure of Saddam Hussein, in other words, the victory is a victory for Iraqis, Shia, and Sunni alike. Moreover, as the Sunni origins of Saddam Hussein are hardly disputable (see Makiya, 2004: 13), the choice can be taken to represent yet another instance of Hussein's duly rewritten (and politically convenient) biography of the country.

In this regard, a last point worth considering is the 'intercommunication' of the Victory Arch with the other Ba'athist monuments in Baghdad. As both Makiya (2004: 19) and Van de Ven (2020: 121) have pointed out, to fully appreciate the politics of Hussein's monumental art, the Victory Arch needs to be situated within the wider urban and architectural complex of its host city, Baghdad. Accordingly, both authors trace the genesis of the arches back to pre-Hussein urban politics in the city, to then put the Victory Arch into communication with other monuments and memorials—in particular with the *Shaheed* and the monument to the 'Unknown Soldier' (Makiya, 2004: 19–32), to which Van de Ven (2020: 121–124) adds the Umm al-Ma'ārik, the Mother of All Battles Mosque.[11] In both cases, the emphasis of the analysis is to uncover how Hussein's search for political authority, his construction of a self-referential cult, as well as his populist-nationalism and open militarism were crystallized through these monuments and thereby inscribed within Baghdad's urban landscape. For the purpose of this chapter, instead, I wish to draw attention to the supplementary work that these other monuments perform to further stabilize the role of the 1988 victory within the country's biography—thereby further amplifying the call for identification it seeks to foster.

First and foremost, then, it is fundamental to note how both the *Shaheed* and the 'Unknown Soldier' are explicitly devoted to—and thereby meant to commemorate—the martyrs of the Iran–Iraq War (see Makiya, 2004: 23–26; and Van de Ven, 2020: 122) in an attempt to lend authenticity and political importance to the conflict and the supposed victory. Moreover, the trope of the martyr (see Makiya, 2004: 23; and Van de Ven, 2020: 113, 129; see also Khoury, 2013) to which both monuments testify and which both seek to reify, foregrounds the redemptive quality of the victory. That is, the rhetoric of the martyr works here as yet another narrational strategy that not only helps

[11] The Mother of All Battles Mosque is another interesting example for the study of victory, given its intended purpose was to 'celebrate the "victory" of Iraq against America in the Persian Gulf War' (Van de Ven, 2020: 122). However, the mosque is not analysed here for both practical reasons and because of the chapter's focus on monuments as material signifiers of victory—i.e., monumental structures designed specifically for the commemoration of military victories, with no other 'explicit' purposes.

construct the victory in the first place but simultaneously justifies the conflict and the deaths it brought about (see O'Driscoll, 2020a: 66–67). Lastly, beneath the monument to the 'Unknown Soldier' is an underground museum that meticulously tells 'the story of Saddam Husain from birth through his early years as a militant ... by analogy with that of the Prophet' (Makiya, 2004: 15), reinforcing once again the idea of Hussein's Shia *and* Sunni origins mentioned above and corroborating the timing function of that identity project.

Overall, both the monument and its associated victory are designed and cast in identitarian terms. Unsurprisingly, then, underpinning their ideation is an explicit intent to reach individuals and elicit identifications *with* and *through* the monument and the victory it materializes.[12] As I have argued, this is visible in both the monumental form of the art, as well as in its repeated (re)mobilizations through public commemorations and celebrations. By treating the Victory Arch as a timing indexical, complementary insights on victory and its politics come to the surface. First, the monument operates as a victory symbol: like the IDF's victory image presented in the book's opening, it seeks to offer proof that the war is over. Way outside the UK military, national culture, and even the Anglosphere, the same dynamics can be observed: victory is constructed as a material event that terminates the war. Second, the victory it seeks to materialize is simultaneously synthesized in a different timing operation, with the Ba'athist Iraq's identity narrative as the timing standard. Here, complementary victory tropes are at work in and around the monument: not only does the arch materialize the victory, but it also acts as a constant reminder of both Iran's supposed inferiority and Saddam's own achievements.

In this latter case, what is particularly salient is the use of victory as a catalyst for identification in the service of nationalism thanks to design choices. In fact, Saddam's Victory Arch might be deemed a quintessential example of a didactic monument: it seeks to 'impar[t] clear, unified messages through figural representation, explicit textual or graphic reference to people, places or events, allegorical figures, and archetypal symbolic forms' (Stevens, Franck, and Fazakerley, 2018: 728). Notably, this need not be the case: some victory

[12] When suggesting that 'the monument functions as a symbol of the identity Saddam wanted his people to rally around in the aftermath of the war', Van de Ven (2020: 124) hints towards a similar conclusion—as does Makiya (2004: 72) when specifying that the monument's 'principal function is to forge this new [i.e., Ba'thist] collective identity supposedly handed down by history'. In other words, the idea of the monument as a device meant to stimulate individual identifications is resonant with the arguments in these contributions. However, both scholars are primarily interested in the political architecture of the monument, which they boldly—and rightly—reconnect to Hussein's nationalist and personal projects, at times even in identitarian terms. In so doing, though, the identity and temporal politics of *victory* are left unattended.

monuments fulfil a similar orientational function even when their message is messy, unclear, and contested. The next two sections turn to two such cases.

## The Bronze Soldier of Tallinn: identity, victory, and narratives

Erected in 1947, the Bronze Soldier of Tallinn was originally titled 'Monument to the Liberators of Tallinn'. Unlike Baghdad's Victory Arch, this monument does not embrace traditional victory tropes, nor does it engage in grandiose displays of a war booty. Instead, the memorial consists of a stone structure at the centre of which is a two-metre 'handsome and melancholic statue' of an 'Estonian soldier in a Red Army uniform' (Kattago, 2009: 149). Indeed, the memorial was originally meant to legitimize Soviet rule in Estonia (Lehti, Jutila, and Jokisipilä, 2008: 397) and could not, therefore, other the vanquished. Following the country's independence in the 1990s, the memorial was renamed 'Monument to the fallen of World War II' in the attempt to de-Sovietize it and accommodate the conflicting interpretation of the Second World War and its aftermaths that the new government supported (Lehti, Jutila, and Jokisipilä, 2008: 398; Smith, 2008: 422). Here, we can already identify two different biographical narratives—and attendant timing attempts—articulated through the monument. Coding them alongside ethnic parameters, Smith (2008) distinguishes one, by Russophone Estonians, as registering 1944 as the year of the liberation of the country from Nazism; and the other, by Nationalist Estonians, reading the event not as 'liberation but simply the replacement of one occupying regime by another' (Smith, 2008: 420).[13] On this account, then, the monument alternatively served as a victorious symbol for one group, and as a less politically charged tribute to the fallen for the other. From a temporal point of view, these narratives also give rise to two different historical chronologies: the first singles out 1944 as the beginning of peacetime and a new chapter of Estonia's national history, while the second codes 1940–1991 as an uninterrupted period of wartime marked by the continued foreign occupation of the country.[14]

[13] Similarly, Kattago (2009: 150–151) has referred to these as '[t]wo memories of the past [that] conflict with one another: an East European (Estonian national narrative) and a Soviet-Russian narrative. If the Estonian narrative emphasizes Estonia as a victim of history and Soviet-Russian aggression, the Soviet-Russian narrative is of Russian victory over barbaric fascism de-linked from the crimes of communism'.

[14] Notably, the orientational capacity of victory is also on display here. Indeed, at the heart of these first two competing narratives is precisely the issue of victory-related situatedness discussed above—that is, acknowledging victory in the Second World War becomes a locus of struggle as it simultaneously moves Estonia alongside the allied powers, together with the USSR, on the victory chariot.

Importantly, commemorations of Victory Day (9 May) by Russian Estonians (Lehti, Jutila, and Jokisipilä, 2008: 398; Smith, 2008: 422)—in the form of public gatherings—kept the first (minoritarian) narrative alive in the country—although contestation or even controversy did not stir until 2004.[15] Indeed, it was only at this point that a plurality of competing and overlapping narratives contextually emerged. This followed the removal of another controversial monument in Lihula, a commemorative stone to 'Estonian men who fought in 1940–1945 against Bolshevism and for the restoration of Estonian independence' (Smith, 2008: 424)[16] featuring 'a relief of an Estonian soldier in German uniform' (Brüggemann and Kasekamp, 2008: 431).[17] In particular, the complex web of narrational enterprises that arose around the Bronze Soldier of Tallinn interlaced multilayered biographies of different collectivities, all structured around the monument.

Capitalizing on the Lihula monument episode, official commentators in Russia (re)launched an (active) timing attempt, promoting a renewed narrative that labelled the Estonian national narrative as fascist (Smith, 2008: 425). Here, readings of 1944 as the beginning of the Soviet occupation, rather than as victory against Nazism, were equated with 'sympathy for "fascism"' (ibid.; see also Brüggemann and Kasekamp, 2008: 432). Moreover, President Putin's official invitation to the sixtieth anniversary of Victory Day celebrations in Moscow (9 May 2005)—extended to the presidents of the three Baltic States—brought both the Lihula episode and the Bronze Soldier back in the public eye (Brüggemann and Kasekamp, 2008: 433–434). Victory Day celebrations at the Bronze Soldier site in Estonia—up until that year quietly unfolding without much resistance or political contestation (Smith, 2008: 422)—attracted unprecedented media coverage by the national press and engendered exceptional political upheavals (Brüggemann and Kasekamp, 2008: 434).[18] When the police drove away an Estonian nationalist protesting the celebration, the

[15] It should be pointed out that the original wooden makeshift of the monument was actually shattered soon after its unveiling, in 1946, as a sign of protest by two Estonian women (Brüggemann and Kasekamp, 2008: 431). However, this act of contestation did not impede the erection of the permanent memorial in 1947.

[16] Brüggemann and Kasekamp (2008: 431–433) specify that the original inscription—changed after the removal of the monument from its original location, in the city of Pärnu, in 2002—read 'To all Estonian soldiers who fell in the second war of liberation and for a free Europe 1940–1945'.

[17] Smith's description (2008: 424) of the same monument reads as follow: 'In flagrant disregard of the taboo in Western Europe against display of Nazi symbols, the stone carried the image of a soldier, machine pistol aloft, wearing the uniform of the Estonian SS Legion'.

[18] Apart from the heated public debate at both the domestic and international levels, discussed below, upheavals included various protests—both for the removal of the Bronze Soldier and for its defence—the most famous of which is probably the Tallinn riot of 26 April 2007 (see Brüggemann and Kasekamp, 2008).

national media shouted at the double standard for the toleration of political symbols (ibid.).

Seizing the moment of heightened hostility, Estonian dissident Tiit Madisson—who was also a crucial actor in the Lihula controversy[19]—led a small demonstration to the monument a couple of weeks after Victory Day. In the process, he contextually announced the birth of a (short-lived) 'Estonian National Movement' (ibid.). On display here is the simultaneous inscription of the Bronze Soldier issue within the birth of a political movement, the mobilization of the Lihula monument episode for the same narrational enterprise through the figure of its leader Madisson, and the broader retracing of Estonia's national biography as championed by the movement. On a similar note, existing political parties in the country came up with their own narratives about the monument, mobilizing the Bronze Soldier as part of their political campaigns for the 2007 elections (Ehala, 2009: 145; see also Brüggemann and Kasekamp, 2008: 435; and Smith, 2008: 426). The Reform Party—which 'present[ed] itself as the embodiment of national coherence and order in the face of a purported external threat' (Smith, 2008: 426)—took a resolute position on the issue of the monument, promising its voters to 'resolve the problem before the next anniversary' (Brüggemann and Kasekamp, 2008: 435). Counting on the vote of the Russian minority, the Centre Party strongly opposed the Reform Party's initiative, decisively splitting the existing governing coalition (ibid.). Interestingly, a new party ran for the elections with a political campaign entirely focused on the Bronze Soldier—the 'ethnically Russian ... Constitutional Party' (ibid.). Having won the elections, Reform Party leader Andrus Ansip kept his word, promptly removing the Bronze Soldier on 27 April 2007 and re-erecting it a few days later in Tallinn's Military Cemetery, where it still rests (Lehti, Jutila, and Jokisipilä, 2008: 393).

Not only is victory's orientational work forcibly evident here, but also its contextual timing function. By taking a stand on one side of the controversy, each of these collectivities went well beyond expressing their political position on the monument. Political alignment through the monument translated into temporal alignment, both in the sense of internal (temporal) stability of a social actor's biography, and external (temporal) stability—or coherence of the actor's role vis-à-vis the broader flow of historical events that actor inhabits. The Reform Party capitalized on the Bronze Soldier to buttress its

[19] During the Lihula episode, Madisson was serving as the chairman of Lihula's municipal council. Openly sympathetic to the project, Madisson actively supported the Association of Freedom Fighters—a group of Estonian veterans fighting under the Germans in 1940–1945—with the erection of the monument in Lihula after its removal from the original site in Pärnu (see Brüggemann and Kasekamp, 2008: 431–433).

privileged position as the true, nationalist Estonian party claiming a coherent national biography—further reinforced by the unveiling of the 'War of Independence Victory Column' a couple of years later (see Brüggemann and Kasekamp, 2008: 438) to explicitly link contemporary Estonia to 1918 independence. The Centre Party's narrative, by contrast, was vocal in embracing historical complexity and a multi-ethnic version of Estonia's past (see Burch and Smith, 2007: 919) in which the two victories—the Second World War and Estonia's first independence—could coexist. As to the two newborn movements, timing is doubly visible, with victory and the Bronze Soldier controversy lying at the heart of both collectivities' actual birth and respective national-biographical narrational endeavours.

Against this background, another series of narratives centring on the monument arose at the international level. A first example is the official position of Russia briefly sketched above. Moreover, several commentators sought to link Putin's heated response to the 'war of monuments' in Estonia to a broader attempt to rewrite Russia's national biography (see Smith, 2008: 425; and Kattago, 2009: 158–159). On this account, victory in the Great Patriotic War becomes the linchpin of historical continuity in today's Russian identity narrative—'the only positive symbol left' to construct the bridge between contemporary Russia and its Soviet past (Kattago, 2009: 159). Accordingly, Russia's strong position and international response are understood in terms of the kind of threat that the Bronze Soldier episode posed—not a mere political dispute over a monument in a foreign country, but a menace to Russia's biographical narrative and ontological security.[20]

Here, it is worth pausing to reflect on yet another case of narrational overlap and competing timing operations: the mobilization of the dispute over the monument in both Russia's biography and the academic debate making the link explicit. On one hand, the politics and ethos of both narrational endeavours are visibly distinct and can be heuristically kept separate, analysed, and juxtaposed. On the other, this dynamic clearly exemplifies a case in which identity narratives and non-biographical ones intersect, proving particularly suited for teasing out the differential insights they offer on victory's orientational capacity. At the national biographical level, the mobilization of the victory in the Great Patriotic War—reaffirmed through the expressed position over the monument—serves as a 'triumphal foundational myth of the new Russia' (Kattago, 2009: 159), explicitly oriented at inscribing the Soviet past within the 'new' country's biography. Here then, the

[20] Here, the flip side of the medal is Russia's patent attempt to (indirectly) influence or *time*, through the articulation of its own biographical narrative, the broader temporal flow it inhabits. Indeed, as suggested above, a victory in 1944 also influences the 'external stability' of the Estonian national narrative.

ways in which victory supplements information about the victor, the vanquished, *and 'the world'* is most visible. This positioning is oriented not only at placing Russia on the victory chariot but also at 'normaliz[ing] the Soviet Communist past into Russian national history' (ibid.) more generally. Importantly, though, similar information about 'the world' is also provided by the parallel deployment of victory in academic re-mediations of the dispute. In this case, the emphasis is on the undisputed assumption of temporal finality—and the capacity to separate wartime from peacetime—that all of the available accounts attribute to victory. Rather than disputing or problematizing this tenet, these accounts seek to reconcile it with—and therefore explain—Russia's political positioning on the Bronze Soldier dispute. When undisputed, however, such an understanding implicitly lends credibility to the very narrative it seeks to examine, uncover, and possibly contest—if anything because the very meaning of military victory overlaps in both narratives. That is, while scholarly accounts claim that the narrative of Soviet-Russia continuity effaces the political violence that followed the Second World War, these accounts also reinforce the notion of victory that enables that narrative in the first place—i.e., they conceive victory as a temporal rupture that ends war and brings peace. As a result, while the narrative is uncovered as nationalist and ideological, the crucial concept upon which it hinges is left untouched.

A second example is found in the narrative surrounding the position of the European Union in response to the Bronze Soldier crisis. Exacerbated by the non-interference of the Kremlin to the blockade of the Estonian embassy by the activist group *Nashi*,[21] the EU officially 'demand[ed] that Russia upholds the Vienna Convention on Diplomatic Relations' (Brüggemann and Kasekamp, 2008: 435). Moreover, Smith (2008: 426) suggested 'the international media showed far greater understanding for the Estonian position', and Brüggemann and Kasekamp (2008: 435) also highlighted the EU 'show of unity' at the EU–Russia Samara summit in May 2007. However, a clear, unitary narrative of the EU on the monument itself is difficult to identify, to the point that even the 'Statement of the European Union on recent developments in Estonian-Russian relations' (OSCE, 2007) mentions the Bronze Soldier but fails to articulate a specific position on its political relevance. Perhaps due to this absence, scholarly analyses of the monument and the memory politics around it emphasize the discrepancy between the European identity narrative and the Estonian nationalist one. Here, the issue is once

[21] Smith (2008: 425) defines the group to be pro-Putin, while Brüggemann and Kasekamp (2008: 436) define it as a 'Kremlin-sponsored patriotic youth organisation'.

again a reading (and attendant temporal implications) of 1944 as the start of a new occupation—sometimes rendered by the idea of the Second World War as ending in the 1990s in Estonia (see Kattago, 2009: 158)—vis-à-vis the centrality of victory in the Second World War, and the associated removal of Nazism, in European history.

For instance, Lehti, Jutila, and Jokisipilä (2008) link the Estonian nationalist narrative around the Bronze Soldier to a broader trend of historical revisionism in former Soviet states. They highlight that 'these tendencies stand in a stark contrast to Western European accounts of the [Second World] war, which are centred on collective mourning for the Holocaust and written *in a deliberately post-national tone*' (Lehti, Jutila, and Jokisipilä, 2008: 396–397, my emphasis). Similarly, Brüggemann and Kasekamp (2008: 436) suggest that conditionality clauses for the accession of Baltic states to NATO and the EU engendered delays in the emergence of collective grievances towards the Soviet past—prioritizing instead the need to come to terms with Nazism as the common European enemy. In a starker account of this same issue, Saarts (paraphrased in Ehala, 2009: 152) argued that 'the Bronze Soldier chain of events was an ethnic counter-reaction to forceful Europeanization in the last decade, when Estonia struggled to meet European standards in multiculturalism and political correctness in order to achieve EU membership'. Here again, I wish to emphasize the deployment of the notion of victory as intrinsically temporal—or as 'always already' an outcome and, therefore, a temporal marker. As in the case of the re-mediation of the Russian biographical narrative above, these accounts assume—and therefore reinforce the idea—that victory is a temporal breaking point in a war/peace binary situation (see Barkawi, 2016). However, in this case, the juxtaposition hinges upon a potential irreconcilability of two narratives positioning victory at different points in time. These, in turn, are deemed significant on account of their 'outward' reverberation, or the influence they exert over the broader temporal flow that these collectivities inhabit.

As Kattago (2009: 160) sought to explain, the importance of the Estonian nationalist narrative sketched out above is perhaps epitomized by key expressions in the country's colloquial language. In the vernacular, the periods spanning from 1918 to 1940—'the years of independence under the First Republic'—and from 1991 onwards are referred to as 'Estonian time' (ibid.). This is often juxtaposed to 'Russian time' or 'Soviet time', which are both used interchangeably—yet another instance of narrational reinforcement, as Kattago notices—as well as 'Czarist time' and 'German time'. All of these times are conceived of, and reified in and through language, as ruptures to the steady unfolding of 'Estonian time'—in turn, a sort of national chronology sanctioned by language. The European version of this same chronology is visibly different: here, victory in the Second World War is the crucial turning point, the creator of a blank slate, and the beginning of a period of unprecedented peace and collegiality in Western Europe. It is by recentring time and chronology—that is, the timing function of victory—that the tension emphasized above emerges. Indeed, on this account, the incompatibility of the two victory narratives associated with the Bronze Soldier is not to be traced to the identity politics of the event, in and for itself, but rather to the politics of its temporal, external reverberations, or the broader time flow that stems from that victory.

On display in this example, then, are clear instances of collectivities including victory as a provider of social orientation within the articulation of their identity narratives. Notably, discrepant interpretations of victory's meaning and temporalities arise from these biographies. For example, the Second World War seems to end at different chronological junctures depending on the actor who is narrating that victory. Moreover, the temporal politics of that specific victory extend beyond the immediate remit of that particular war or conflict. To celebrate a victory in the Second World War in the 1940s, for instance, is to overshadow that the following fifty years were experienced by some as a military occupation by a foreign regime.

The Bronze Soldier of Tallinn, however, also shows that identification and social orientation flow from and via victory even in the absence of the clear messages and explicit visual references that are proper of didactic monuments. In fact, it is precisely through contestation and controversy that different identity narratives came to be articulated that, simultaneously, sought to shape the temporal contours of victory. This point is doubly important. First, the example shows that even biographical narratives which synthesize victory within their storyline are constitutive of victory. Second, it exposes the unstable nature of victory narratives *and* symbols, even when the symbol is a monument erected precisely to pin down a specific victory narrative. Indeed, not only is victory a temporal, sense-making device the contours of which are constantly negotiated, but monuments are unstable too—or, as Koselleck (2002: 288) famously put it, 'over the course of time ... the intended identity similarly eludes the control of those who established the memorial. More than anything else, memorials erected permanently testify to transitoriness.'

The following, final section turns to an example in which contestation is brought one step further, in the direction of institutionalized counter-monumental practices that explicitly seek to reappropriate the meaning of a victory monument. While in the case of the Bronze soldier of Tallinn the reproduction of the hegemonic understanding of victory and its relationship to time might be attributed to the relative success of certain identity narratives over others, in this last case, a univocal interpretation of the victory monument gets institutionalized in the wake of multilayered contestation. And even then, the link between victory and temporality remains intact.

## Bolzano's Victory Gate: counter-monuments and hegemonic victory

Erected in 1928 at the direct orders of Benito Mussolini, the *Monumento alla Vittoria* (Victory Monument) or Victory Gate in Bolzano, is considered 'the first built monument of the fascist era' (Favargiotti, Busana, and Cappelletti, 2020: 2010, fn13) and the 'first truly fascist monument' (Schnapp, 2020: 533; see also Contested Histories, 2021: 4). The monument is a Roman triumphal arch measuring a huge '19 meters wide, 20.5 meters high, and 8 meters deep', sustained by fourteen fasces-shaped columns (Steinacher, 2013: 652).[22] Its express purpose was to celebrate the Italian victory in the First World War (Carlà and Mitterhofer, 2017).

Just like Baghdad's Hands of Victory, the Victory Gate in Bolzano fulfilled this purpose *by design*. In addition to the classical Roman triumphal arch shape, indeed, the monument presents several elements that signify victory. On top of the main façade is a high relief sculpture of a Winged Victory—*Vittoria Alata* or *Vittoria Sagittaria*—about to fire an arrow in the direction of Austria (see Steinacher, 2013: 652; Carlà and Mitterhofer, 2017; Hökerberg, 2017: 762; Di Michele, 2020: 154; Contested Histories, 2021: 5; for a close-up view of this relief see Soragni, 1993: 36–37).[23] The goddess stands above a Latin engraving stating: 'Here at the border of the fatherland stands a marker. From this point on, we educated the others with language, law, and culture' (see Steinacher, 2013: 652; see also Hökerberg, 2017: 762; Contested Histories, 2021: 5).[24] Not only does the statue work as a victory symbol—an embodiment of victory—but, together with the relief, it also puts

[22] Fasces, or *fasci littori*, are the quintessential symbol of Italian Fascism—to the point of inspiring the movement's name. Visually, *fasci littori* are bundles of rods holding an axe at the centre. The fourteen columns that comprise the Victory Arch in Bolzano, then, are but massive, three-dimensional sculptures of this symbol. For a close-up of the columns see Soragni (1993: 29). See also fn. 26.

[23] Drawing another parallel with Saddam's Arch, the arrow of the Winged Victory on the monument in Bolzano points towards Austria, just as the tips of the swords in Baghdad point towards Israel and Iran.

[24] The Latin inscription reads: 'Hic Patriae Fines Siste Signa Hinc Ceteros Excoluimus Lingua Legibus Artibus'. According to Hökerberg (2017: 763), '[a]n earlier version of the text had included the word *barbaros*, but this was replaced by the less provocative *ceteros* (the others), to mitigate the insult against those German speakers who were now Italian citizens'.

in place traditional othering processes analogous to those instantiated by Saddam's arch against Iranians. For example, the monument was purposefully built on top of the remnants of an unfinished, Austrian memorial to the *Kaiserjäger*—an Austrian regiment that fell fighting the Italians during the First World War—using materials obtained from its demolition (Contested Histories, 2021: 4–5; see also Soragni, 1993: 17–18, 25; and Steinecher, 2013: 652). 'The appropriation of the site of the *Kaiserjäger* war memorial', Hökerberg (2017: 762) explains, 'was a manifestation of the newly achieved Italian supremacy in South Tyrol, including that over its German-speaking inhabitants'. To put it bluntly, the new Victory Arch sought to materially and metaphorically phagocytize the older, unfinished Austrian memorial.[25] Embedded *in* the monument, then, is also the oppositional logic that is proper of the hegemonic understanding of victory.

Notably, the monument also worked as a vehicle for fascist propaganda. In fact, the Victory Gate can be deemed polysemic in character, embodying at once victory in the First World War *and* Italian Fascism. For scholars working on the Contested Histories (2021: 2) project, the latter aim was even more prominent than the former. They write: '[w]hile officially dedicated to Italy's triumph and sacrifice during the First World War, the Victory Monument symbolized Fascist Italy's dominion over Bolzano's German-speaking population and formed a part of the attempts to Italianize the city' (ibid.). Once again, several aspects of the monument, including iconography, design, and location corroborate this interpretation. The fourteen fasces-shaped columns that compose the monument, for instance, were invented ad hoc by the Gate's architect, Marcello Piacentini, precisely to give a fascist flavour to the classical victory arch from Ancient Rome that inspired the Gate's design (see Hökerberg, 2017: 762).[26] Similarly, the southern side of the entablature celebrated Mussolini and fascism with the inscription 'Ben. Mussolini, Ital. Duce A. VI'. (Hökerberg, 2017: 762), where the latter part stands for 'year 6' to

[25] According to Schnapp (2020: 535) 'granite remains [of the *Kaiserjäger*] were used to reinforce the new monument's foundation and to line its crypt . . . Other instances of recycling would include the melting down of over 200 Austrian coins to forge the trowel used in the opening ceremonies'. For Steinacher (2013: 652), 'Mussolini's arch literally stood on the ruins of the Habsburg monarchy, symbolizing Italy's rule over South Tyrol and its claim on the new border in the Brenner Valley'.

[26] According to Hökerberg (2017: 761), 'Piacentini not only conceived the Monument to Victory as the true fascist monument, but he also declared it "the architectural imprint of the fascist spirit", dedicated to the fascist party, at its inauguration in 1928'. The adoption of this, newly designed fasces-shaped columns, indeed, sought to 'mar[k] the debut of a new columnar order: the "lictorial" column, the Mussolinian addition to the classical orders codified by Barozzi da Vignola—Doric, Ionic, Corinthian, Tuscan, and composite' (Schnapp, 2020: 534). According to Soragni (1993: 31, my translation) the launch of the fasces-shaped column was a 'strongly propagandistic' move 'to materialise an ever-evolving "fascist modernity"'. Unsurprisingly, the 'lictorial column' would figure time and again on other fascist monuments throughout the country during the 1930s (ibid.; see also Hökerberg, 2017: 762).

mark the date of the monument's inauguration on the fascist calendar—i.e., 1928, six years after the March on Rome in 1922.[27] Moreover, the monument was strategically erected in *Piazza alla Vittoria* (Victory Square) to separate 'the pre-existing German historic centre in the east and the new Italian neighbourhoods in the west' (Contested Histories, 2021: 3; see also Soragni, 1993: 1–4).[28]

Unsurprisingly, the *Monumento alla Vittoria* has been the cause of much political turmoil from its unveiling until the present day. 'Of all the Fascist monuments and buildings in South Tyrol', writes Steinacher (2013: 655), 'the Victory Monument has stoked the greatest tensions, particularly within the last thirty years, which have seen an escalation in polemics, protests, and clashes. From its earliest days, the arch has been a barometer of local ethnic frictions'. On the same day of the inauguration, in 1928, for example, a protest was held in Innsbruck, Austria to contest its erection (Contested Histories, 2021: 6). In 1943, a Nazi group vandalized parts of the monument (Carlà and Mitterhofer, 2017: 20), which was then fully restored by 1949 thanks to Italian veterans' fundraisers and institutional aid from the Italian state (Di Michele, 2020: 156). Political demonstrations against the monument have also taken place routinely, especially on occasion of victory celebrations[29] organized at the Victory Gate up until 1997 (Carlà and Mitterhofer, 2017: 20). Petitions for the monument's demolition brought together groups across the political spectrum, including 'regionalist forces, pacifists, and even Communists' (Schnapp, 2020: 536). In the late 1970s, Tyrolean nationalists attempted to bomb the monument (Steinacher, 2013: 656; Schnapp, 2020: 536; Contested Histories, 2021: 6–7). As a response, the Arch was fenced in, and surveillance heightened (ibid.). In short, and not unlike the Bronze Soldier of Tallinn, Bolzano's Victory Monument sat at the centre of a wealth of identity narratives, at various levels. Also in this case, the monument spurred political contestation across factions and worked as a provider of social orientation.[30]

[27] This dedication was removed in the aftermaths of the Second World War (Schnapp, 2020: 536). Of note, however, is the patent, active timing attempt to reconsider the Gregorian Calendar and champion the adoption of a new, fascist one.

[28] For Andrea di Michele (2020: 155, my translation) the monument sought to 'reiterate the solidity of the new border ... and offer a visual representation of the change of sovereign and the imposition of Italianness'. According to several scholars, indeed, the erection of the Victory Monument in Bolzano should also be read within the context of a fascist, urban planning project that was part of the abovementioned plan to 'Italianize' Bolzano. For a discussion of the urban role of the monument, as well as other initiatives through which the 'Italianization' of Bolzano was to be achieved see Soragni (1993), Steinacher (2013), and Favarigiotti, Busana, and Cappelletti (2020).

[29] Interestingly, these celebrations took place on 4 November, on 'the anniversary of the end of World War I in 1918 (for Italy) and of Italy's victory over Austria-Hungary' (Steinacher, 2013: 655). Of note, here, is the discrepancy in date with, for instance, the UK (see Chapter 3).

[30] For a detailed account of the controversies and the positioning of specific political groups see Steinacher (2013).

The remainder of this chapter, however, does not focus on the similarities between the victory monuments in Baghdad, Tallinn, and Bolzano. I draw out, instead, two key differences. First, political controversies in Bolzano ultimately led to institutionalized, counter-monumental practices that have been repeatedly praised for the ways in which they managed to deal with the monument and its violent legacies. The monument, in other words, was not 'neutralized' through relocation, like in Tallinn, but kept on its original site and modified for the sake of offering historical contextualization and generating critical engagements from the public. Second, contestation *and* institutional interventions did not centre solely, or even mainly, around the issue of victory. They focused, rather, on fascism. As a result, the victory side of the monument's signification was left intact. Notably, while these counter-monumental practices were not immediately concerned with victory, they paradoxically ended up contributing to the reproduction of victory's traditional temporal parameters. The *Monumento alla Vittoria*, in other words, demonstrates the solidity and persistency of hegemonic understandings of victory on a global scale, even in the wake of full-on contestation and institutionalized counter-monumental practices.

Following the intensification of political disputes in the late 1990s and early 2000s, a proposal was accepted in 2011 to 'develop a documentation centre in the Victory Monument's crypt' that would 'historicize the monument and defuse its fascist connotations' (Contested Histories, 2021: 8). According to the commission of experts in charge of the project, 'an appropriate and knowledgeable work of contextualisation, anchored in the most up to date historiographic debate, could finally allow the contrasts that originated with the presence of a political and ideological symbol of the "century of the extremes to be overcome"' (Commissione, cited in Angelucci and Kerschbamer, 2017: 63). The result was 'BZ '18–'45: One Monument. One Town. Two Dictatorships', a permanent, onsite exhibition opened to the public in 2014 (see BZ '18–'45, n.d.). There are several elements to this exhibition. From the outside in: a LED ring on one of the columns, displaying the title of the exhibition in Italian, German, and English; a sound installation with visual projections of quotes from Hannah Arendt, Bertolt Brecht, and Thomas Paine in the atrium and crypt, where word clouds of the most recurrent themes of the exhibitions have also been painted on the wall, in light grey shades, on the space where floral wreaths used to be hung; and three different curatorial pathways that the visitor can approach simultaneously or independently (see BZ '18–'45, n.d.; gruppe gut 2014; see also Angelucci and Kerschbamer, 2017; Carlà and Mitterhofer, 2017; Hökerberg, 2017; Di Michele, 2020; Schnapp, 2020).

The ring works as 'highly visible intervention which breaks the monumental nature of façade ... "neutralizing" and "contradicting" the ideology of the monument without damaging or erasing it' (Carlà and Mitterhofer, 2017: 25). The sound and visual installation, in turn, seek to instantiate a dialectic exchange between the fascist, physical inscriptions on the wall and the anti-dictatorial (and antifascist) quotes projected onto them, with the aim of inspiring visitors to engage critically with the monument, its legacies and values (Schnapp, 2020: 540). Finally, the curatorial pathways offer three overlapping, different historical narratives: the external wall recounts the history of the monument, the internal wall that of Bolzano (city and province) under fascism and Nazism (from 1918 until 1945), and four corner rooms explore metanarrational questions on monuments, memorialization, and the design and architect of the Victory Gate (Contested Histories, 2021: 9).

'In their questioning of the original architecture', write Angelucci and Kerschbamer (2017: 65), 'these interventions constitute themselves as counter-monumental: they are provocations and invitations to interaction and engagement with the burden of memory' (see also Carlà and Mitterhofer, 2017: 26; Hökerberg, 2017: 769–772). According to Young (1992: 271), counter-monuments are 'brazen, painfully self-conscious memorial spaces conceived to challenge the very premises of their being'. 'BZ '18–'45' seems to comply with Young's definition particularly well, especially when it comes to challenging fascism. The curatorial pathway on the internal wall of the crypt, for example, 'describes how the city was affected by the two dictatorships', homing in on selected themes, including 'the fascist repression, the discrimination of the German-speaking local population, the Italianisation process, and terrors of the Nazi occupation, such as the Bolzano concentration camp and the deportation of Jews' (Hökerberg, 2017: 767). In so doing, the exhibition has not only redefined the meaning and value of the Victory Gate, but also reconfigured some of the social dynamics at the heart of its political contestation. According to heritage and architecture scholar Håkan Hökerberg (2017: 768–769; see also 771–772), indeed

> new non-material dimensions have been added to the monument. These [counter-monumental] processes have made a reinterpretation possible: the exhibition and its interpretation of history recognise the controversial character of the monument, but identify it as a product of past (and thus no longer valid) political circumstances. This departure from the original understanding of the monument, by highlighting the historical failures with which it has been connected, offers a new interpretation that no longer divides the population of Bolzano, but conveys

> a message of reconciliation. Historicisation does not in itself cause reconciliation; it could also, by contrast, be used for the purpose of segregation. It was the decision to use historicization against fascist rhetoric that unlocked the potential to resolve the dissonances associated to the monument.

Notably, the exhibition has been repeatedly praised as 'a great success' by practitioners and the public. For example, 'BZ '18–'45' received a special recommendation at the European Museum of the Year Award in 2016 as a 'highly courageous and professional initiative [that] promote[s] humanism, tolerance and democracy' (Hökerberg, 2017: 767; see also Guarnieri, 2017: 24). As scholars from the Contested Histories (2021: 8) project noted, it 'received a generally positive response from both the German and Italian-speaking press'. Moreover, since the inauguration of the exhibition, 'controversies surrounding the monument have largely disappeared' (Carlà and Mitterhofer, 2017: 26; cf. with Budabin, 2023).

On the one hand, this is unsurprising. First, the focus on 'desacralising' (see Hökerberg, 2017: 768–769) the monument and contextualizing its fascist nature and origins is perfectly in line with the type of political contestation that it attracted over the years. Counter-monumental practices, in other words, focused predominantly on fascism (as opposed to victory) because it was the monument's signification as a fascist symbol that ignited the political disputes around it. Second, the birth of the exhibition 'was rooted in a call from Bolzano's *Archivio Storico* in May 2010 to historicise the monument and *defuse its fascist connotations*' (Contested Histories, 2021: 8, my emphasis; see also Hökerberg, 2017: 760). Similarly, another appeal often credited with kick-starting the counter-monumental practices that led to the exhibition is a petition from a group of historians, 'dated February 4, 2011, and titled "Let's Resolve the Problem of Fascist Monuments Together"' (Schnapp, 2020: 536).

On the other hand, that counter-monumental practices focused exclusively on fascism can be deemed a peculiarly selective choice. As suggested earlier, indeed, the *Monumento alla Vittoria* is polysemic in nature: it simultaneously signifies victory *and* fascism. While effective, the institutional historicization and contextualization processes carried out through 'BZ '18–'45' leave the 'victory side' of the monument virtually intact. However, as this book has so far demonstrated, victory is not a neutral, apolitical notion. Victory in the First World War, for instance, played a fundamental role within the ideological and biographical apparatus of Italian Fascism. In fact, some scholars explicitly read the erection of the Bolzano Victory Monument as yet another 'symbol of the fascist appropriation of the First World War legacy' (Hökerberg, 2017: 759; see also Contested History, 2021). 'Fascism', Pisani

(2017: 9) explains, 'had always presented itself as heir to the Great War. Once it came to power, it undertook a series of initiatives aimed at establishing a monopoly on commemorative monuments and memorials to the fallen'. According to Hökerberg (2017: 761), '[t]he legacy of the war became an integrated part of fascist ideology: worship of the nation, of patriotism and of victory' (see also Di Michele, 2020: 150).

In this sense, the lack of engagement with the 'victory side' of the monument might be read, in the first instance, as a missed opportunity to further discuss or contextualize, *via* victory, the fascist nature of the monument. Most importantly, at least for the purpose of this book, it risks lending credence and reproducing the hegemonic understanding of victory discussed and challenged so far. The very title of the exhibition, for instance, reinforces the idea that military victories are clearly identifiable *events* with specific temporal coordinates—in this case both World Wars, in 1918 and 1945. This is especially clear from the official documents that outline the architectural-historical intervention of 'BZ '18–'45': '[F]or the logo ... [w]e opted for a typographical solution that includes the abbreviation of Bolzano "BZ" and the final digits of the two world wars, 1918 and 1945' (gruppe gut 2014: 13; my translation). The exhibition's title, in other words, works as a timing operation: it *times* the different narratives on display *through victory*. In so doing, however, it also reproduces these specific victories' temporal coordinates, the idea that both wars ended in those years, and a broader understanding that victory does bring war to an end.

On a similar note, no specific neutralization or contextualization strategy has been adopted for victory symbols. For example, the word clouds in the atrium have been painted 'in the shape of victory-wreaths, only with keywords in Italian, German, and English rather than leaves' (Di Michele, 2020: 158, my translation). Analogously, one of the panels in the inner-wall pathway dedicated to explaining the reconciliation purpose of the exhibition is titled 'A monument to other victories?' (Angelucci and Kerschbamer, 2017: 70). Also in these cases, then, the idea of challenging the nature of the monument does not seem to apply to the notion of victory, but 'only' to fascism. As a result, the Victory Gate can now be deemed a counter-monument, perhaps even a new symbol of 'a modern, democratic context' (Hökerberg, 2017: 760) or a 'monument to the victory of democracy' (Angelucci and Kerschbamer, 2017: 65), but a victory monument all the same. Indeed, in praising the exhibition and its counter-monumental value, a few observers have suggested that the Victory Gate no longer 'represent[s] "that victory" but the victory of "the new South Tyrolean". In other words, it "is already in substance a monument of peace"' (Carlà and Mitterhofer, 2017: 31).

Obviously, 'BZ '18–'45' has not been completely immune to criticism. Malcolm Angelucci and Stefano Kerschbamer (2017: 71), for example, have challenged the counter-monumental value of 'BZ '18–'45', arguing that the exhibition 'attempt[s] to propose a new hegemonic message whilst embracing many of the original, fascist strategies: from mottoes to a theatrical use of space, from performative arrangements to monologic, myth-building messages', thereby 'investing in a new telos that grounds the democratic town in the same soil as the fascist one'. Arguably, the lack of engagement with the monument as a victory symbol might be added to this line of critique. A broader discussion on the politics of counter-monuments or even historiographical practice, however, is clearly beyond the scope of this book. Instead, I wish to suggest that there is a lesson to be learned, here, relating to the subtle, complex dynamics of reification of the idea of victory. Indeed, on account of the polysemic nature of the Victory Gate, even particularly successful counter-monumental practices ended up leaving victory's oppositional logic and temporal contours intact. Read in conjunction with the vicissitudes surrounding Baghdad's Hands of Victory and the Bronze Soldier of Tallinn, the example of the *Monumento alla Vittoria* in Bolzano further highlights that hegemonic understandings of victory are hyper pervasive, intricate, and particularly resilient.

## Conclusion

The production of monumental art in the wake of armed conflicts has been a frequently recurring practice throughout history. Suggesting it is also a widespread victory practice, this chapter has shed additional light on the link between victory and time that sits at the heart of this book. First, I argued that victory often works as a situational device: it provides information about the spatio-temporal and sociopolitical location of a given subject. It is for this reason that victory utterances are not restricted to the realm of warfare, extend well beyond the wartime battlefield, and figure frequently in identity narratives. Second, I showed how artefacts such as victory monuments can contribute to the material production of victory alongside, and even notwithstanding, military developments. Indeed, as was the case with Baghdad's Hands of Victory, monumental art can produce 'proof' of a given victory amid ongoing hostilities and in the absence of a clear military record.

This example, then, contextually showed how the victories that social subjects incorporate in their biographies can well be mythical, or even staged, like in the Bint Jbeil example that opened this book. According to Duncan

Bell (2003: 75), a (national identity) myth is 'a story that simplifies, dramatizes and selectively narrates the story of a nation's past and its place in the world, its historical eschatology: a story that elucidates its contemporary meaning through (re)constructing its past. ... Myths are constructed, they are shaped, whether by deliberate manipulation and intentional action, or perhaps through the particular resonance of works of literature and art'. Saddam's Victory Arch and the accompanying narratives on victory in the Iran–Iraq War seem to fit Bell's definition exceptionally well. Indeed, not only did the monument predate the military success it sought to celebrate, but it also gave material form to a very specific narrative of Iraq's past. To this day, the outcome of that war is disputed, and the monument can be read as an attempt to help that victory happen by inscribing a version of history in which Iraq did win that war into the public space.

Although it constitutes this kind of material proof, a victory monument—as is also the case with 'facts on the ground'—is not irrefutable nor immutable. As the examples of the Bronze Soldier and the Victory Gate in Bolzano demonstrate, narratives of contestation might try to change a monument's meaning and with it the victory it signifies. There are at least two key take-aways from these examples. From a theoretical standpoint, the making of victory has no single temporal positioning: practices implicated in the social construction of a specific victory and its temporal contours can take place *post-conflict* but also *in medias res*—or, to put it bluntly, at any given point in time—and even change those temporal coordinates. From a methodological standpoint, the same victory symbol gives access to overlapping, and even competing, timing attempts—just like the tattoos analysed in the previous chapter. Notably, these attempts can take various forms. More or less direct narratives of a given victory might be proposed, but equally even extratextual means may be employed, such as the deliberate choice of certain materials for the construction of an artefact.

# Conclusion

On Tuesday, 8 August 2006 paratrooper brigade commander Hagai Mordechai is sent back to Bint Jbeil with the order of capturing a compound and 'rais[ing] the Israeli flag in the locality's headquarters' (Harel and Issacharoff, 2008: 175; see also 176)—i.e., the balcony from which Nasrallah made his spider's web speech in May 2000. Met by concentrated enemy fire, Mordechai's already exhausted forces suffer two casualties, but successfully gain control of the junction (Harel and Issacharoff, 2008: 176). The commander deems the flag objective minor and not worth risking other lives. He decides, instead, to get a videorecording of the flag 'flying from an adjacent building' and exit Bint Jbeil (ibid.). 'Flying the flag was not the main thing. I had no intention of endangering the lives of my men for it' (Mordechai, cited in ibid.). The result, according to Yoav Galai (2019: 302) is far from satisfactory:

> The video was shot from another rooftop. The flag-raising soldier is distant, grainy and barely visible. As he steps onto the balcony, the commander orders him to crouch, for fear of fire. The cameraman taking the video is then asked to zoom out and take a wider shot, as proof that they are in Bint-Jbeil, but then the flag is hardly visible. The Israeli soldier was raising the flag to demonstrate that the Hezbollah stronghold had yielded. However, the narrow angle and the blurry image give away the great distance between the flag-raising soldier and the camera. ... Together with the accompanying 'director's commentary' the performance appears more like an act for the camera: the momentary presence of a body enacting its dominance over the landscape while not exposing itself to it, in stark contrast to the Hezbollah images, both of Nasrallah and of the flag planting in Nabatieh, and completely in line with the idea of 'victory staging'.

The clunky result could hardly convey the sense of grandiosity and victoriousness that the IDF was after. For Eyal Zisser (2010: 41), indeed, the footage was supposed to provide an image akin to 'the iconic photograph of the American marines raising the flag at Iwo Jima'. A blurry video of a crouching soldier waving a hardly visible flag, far away from the camera, was

*Military Victory Beyond the Battlefield*. Mirko Palestrino, Oxford University Press. © Mirko Palestrino (2025).
DOI: 10.1093/9780198955818.003.0007

surely incomparable. In addition to the unpleasant aesthetics, Galai (2019: 303) suggests, the IDF soon realized that the casualties suffered during the mission had probably been the result of friendly fire—yet another reason to discard the image. Overall, for him, the footage 'told the same story of failure that the Winograd report did' (ibid.)—that is, one of military failure and not success. The Bint Jbeil footage was not able to mark the end of the 2006 Lebanon War. Notably, as Harel and Issacharoff (2008: 176) point out, it did not stop '[t]he quest for a victory picture' either. According to Galai (2019: 303), a victory image was found just a few days later. First published in newspapers on Friday, 11 August, 'with the caption "I am victorious"', the picture shows a close-up of an injured soldier, Major Tomer Bohadana, 'holding up two fingers to flash the victory sign' while receiving medical assistance (ibid.). Arguably not decisive in terms of bringing the conflict to closure (see Zisser, 2010: 41), the image was immediately amplified by the Israeli media and was well received among the Israeli public (Galai, 2019: 304, 2021). 'Ultimately', Zisser (2010: 41) writes, 'Israel remained without an Iwo Jima kind of "picture of victory"'. In this context, it is beside the point whether a picture embodying victory and published in a newspaper could have conveyed a meaningful message about whether Israel had indeed achieved anything in the war, or not'.

Not only does Galai (2019, 2021) disagree on the absence of a 'victory image', but he also argues that Bohadana's V-sign picture 'assured the public that the positive values ascribed to warfighting were being upheld' (Galai, 2019: 303), and 'was used [as] a distillation of the idea of victory into a simple gesture, a reminder that the fighting spirit of the Israeli soldier survives his injuries, that the journey into battle is noble and that the return is heroic' (Galai, 2019: 304). In short, the image was successfully deployed to corroborate nationalistic narratives of heroism and victory, garnering attention *and* support towards the IDF and its subsequent missions. Indeed, the victory image was so impactful—at least as an idea or trope—that it was, and still is, retained as a formal military objective in several IDF military operations after Lebanon 2006 (Galai, 2021: 211). At a minimum, then, the IDF's continued quest for 'victory images' acts as a tangible reminder that questions of victory and associated practices are not 'beside the point' (Zisser, 2010: 41). In fact, Bint Jbeil and subsequent operations aimed at capturing a picture of victory—visually or otherwise (see Galai, 2021)—offer a vivid example of the co-constitutive relationship between victory's narrational and material dimensions. 'Photography', Galai (2019: 304) writes, 'was written into the battle plan'. Just like UK doctrinarians recommend fighting narratives or taking the media timetable into account when planning and carrying out military actions, the IDF designed an entire operation aimed exclusively at collecting—or producing—evidence that could back up their victory narrative. Upon failing, the IDF retimed their narrative, adducing new material 'proofs' in support of a biographical storyline marked by tenacity, victoriousness, and heroism.

Read against the backdrop of this book's arguments and findings, Web of Steel fails to surprise or prove exceptional. Rather, the battle of Bint Jbeil and associated video become just another item in a long list of victory practices that straddle the lines between identity and military developments, wartime and peacetime, battlefields and peace zones, strategy and politics. They also illustrate vividly the violence carried out by and through apparently non-war-like victory practices. A sense-making device that seeks to order and synthesize varied phenomena and events with the ultimate aim of convincing audiences, victory brings conflict to closure only if fruitfully construed as the end of war. Photographs, videotapes, images, and other material artefacts might well aid the emplotment of these narratives—even when manufactured ad hoc or mobilized to back up mythical stories. Inherently unstable, however, these material traces of victory might just as well spur contestation, aid antithetical accounts, and work as privileged entry points—or 'scars of victory' (Campbell and Steele, 2017)—into the narratives they support for the sake of uncovering, resisting, or even rewriting them. Indeed, even successful symbols and narrational enterprises might not sit well with the experience of some individuals or might hinge on their continued reproduction—long after the war to which they bring closure is supposedly over. Whether working towards the construction of a proleptic victory, *in medias res*, or acting on a given conflict's story retroactively, symbols and narratives of victory are but temporal operations focused on offering a politically convenient interpretation of war.

## Timing victory: the road ahead

This book has argued that we should not only redefine victory as a sense-making device, but we should contextually rethink how we approach its study. Dichotomies that are normally associated with victory should be treated with a certain degree of scepticism. In fact, they should be taken up as precious occasions to shed light on the temporal politics enacted through a seemingly self-explanatory notion. The examination of weaponry, tactics, strategy, diplomatic, and economic efforts to bring wars to a successful conclusion ought to be complemented by—and carried out alongside the study of—individual war stories, commemoration, and other sense-making practices from celebrations to history teaching, doctrine making, or even videogaming. Materiality and narrativity, it was shown, are more intermingled than one might usually think. If victory is made and remade inside and outside the battlefield, during both wartime and peacetime—in fact, if victory is constitutive of these ideas—then victory research should also transcend these distinctions.

In doing precisely this, the book has offered three main contributions to IR, War Studies, and germane disciplines. First, it offered a new understanding of victory, not as an outcome of conflicts, but as a temporal marker: a concept made *in* and *through* time. Second, it designed and implemented a novel method for the study of concepts, symbols, and narratives that blends theory and practice, textual and extratextual sources, analysis and meta-analysis. Third, it furthered the theory and study of time and temporality, extending timing theory in at least two substantial ways. In the remainder of this concluding chapter, I expand on each of these interventions, outlining both their political relevance to contemporary global politics and implications for future research.

## IR and the critical study of war

As scholarship on victory and its meaning experiences a new resurgence (see, for instance, O'Driscoll, 2019, 2020a, 2020b; Hall, 2022; Hartnett et al., 2022; Hom and Campbell, 2022b), this book advanced a timely, novel conceptualization of military victory that solves some long-standing conundrums at the heart of IR and War Studies, while simultaneously opening both fields up to complementary research agendas. Grounding my investigation within the recent traditions of Critical War and Military Studies, I offered a new ontology of victory as a temporal, sense-making device co-constituted materially and narrationally through multiple and overlapping timing processes. Victory, on this account, is not one of the outcomes of armed conflicts, nor the material result of military developments 'on the ground', but it is often construed as such. Seemingly self-explanatory, victory is instead the product of knowledge production processes—or timing operations (Hom, 2018a, 2020)—in which a plethora of phenomena, events, and people are synthesized, ordered, and coordinated in a coherent way for the sake of making war intelligible, convincing audiences, or furthering one's political agenda.

Such an understanding of victory offers some explanations, for instance, as to the widely recognized fuzziness of the notion, and to the otherwise almost inexplicable affective pull that victory seems to exert on enlistees or crowds when uttered. In terms of the former, I suggested that victory is constantly caught up in the middle of narrational enterprises that seek to establish, reinforce, or disrupt its conceptual contours. In the wake of unfavourable or indecisive military developments, the IDF at Bint Jbeil and Saddam Hussein in the Iran–Iraq War sought to manufacture artefacts that could alter or stir those same 'facts' in their favour. With the aim of garnering support from the American public, both Bush and Obama called for victory in Afghanistan—the former to extend the conflict, the latter to bring it to an end. Similar dynamics lie at the basis of victory's capacity to appeal to multitudes or inflame their spirit—a well-known empirical observation that we often struggle to explain. Among the temporal operations performed through victory, I argued, is the provision of social orientations to both individuals and collective subjects. Mobilized, for example, in the articulation of identity narratives, victory positions the Self sociopolitically, subtly supplementing information about a subject's relationship to friendly others and alien foes or inscribing their story within broader historical trajectories. In turn, the affective politics of these operations are unleashed, as individuals identify with victorious collectivities, take distance from antagonists, break with an undesirable past, embark on the path towards a more enticing future, or subscribe to ancillary biographical traits.

Besides offering some potential answers, this approach to victory sought to also raise complementary and new questions—some of which seem particularly timely in light of recent world-political events. As the pages of newspapers from across the world get filled with reflections on the US defeat in Afghanistan (see, for instance, Bishara, 2021; Coll, 2021; Jacques, 2021; Willasey-Wilsey, 2021), the implications of the Taliban's victory (The Guardian, 2022), or calls to get ready 'for a Ukrainian Victory' (Applebaum, 2022), research on victory practices appears to be especially urgent. What are the politics of these narrational enterprises? How do they interact with facts on the ground, the war experience of those who are living these conflicts, or even global politics in general? Are these additional proofs of what Matthew Ford and Andrew Hoskins (2022) call the advent of 'radical war'? For these two authors, indeed, contemporary conflicts are characterized by a new ecology of war in which data and information are as deadly as tanks, drones, rifles, and other weapons (ibid.). This book has shown that narratives and symbols are not necessarily non-violent, and that failing to understand how they interlace with, for example, the planning and execution of military operations leads to a problematic sanitization of non-immediately martial processes. Clearly, the politics of war narratives are not limited to war endings or the ideas of victory and defeat. On a similar note, the co-constitutive relationship between the materialities of warfare and their narration is likely to extend beyond the narrow remit of victory and its temporal politics. Thomas Gregory (2022: 479), for example, has recently shown that the rise in popularity of body count metrics—and ancillary practices, such as condolence payments (Gregory, 2020)—helped 'enable and enhance the effectiveness of military operations' in Afghanistan, and should be deemed 'complicit in the violence experienced by Afghan civilians'. What other apparently peaceful or non-immediately martial processes are similarly implicated in the perpetration of violence? How exactly do these practices shape war, its preparation, and its aftermaths?

This book has also shown that the deployment of supposedly innocuous temporal categories—such as wartime and peacetime—is, in fact, riddled with fundamental, sociopolitical consequences. The re-mediation of chronic war experiences within orderly recounts of temporally bounded wars, for instance, can lead to the effacement of these war stories and the marginalization of certain political subjects. Indeed, temporal references—or timing indexicals (Hom, 2018a, 2020)—provide access to timing processes through which war is made intelligible and as part of which effacements, erasures, and other political moves necessarily take place. In which other ways, beyond victory and defeat, is war timed? Which exclusions do they underpin, and which phenomena, events, and people do they foreground? Are different temporal dynamics unleashed through them? What are their sociopolitical consequences? Which subjects do they 'cast into motion' (Barkawi and Brighton, 2011: 136)? As this text argued, wartime and peacetime politics travel both realms and blur their boundaries, calling for research that disputes and traverse clear demarcations between war's temporal phases. It seems pertinent, therefore, to question research endeavours that simply assume, and practices that reproduce, similar distinctions and fail to engage with their social constructedness and exclusionary politics. Traditional normative enquiry into the ethics of war is a prominent example of the former, while curriculum design and teaching are obvious instances of the latter. How would *jus in bello* and *jus post bellum* principles change if we started from the assumption that there are no clear wartimes and peacetimes? Should we rethink transitional justice in light of the political and analytical pitfalls of the war/peace(time) binary? How can we teach history in a way that does not risk effacing marginalized lived experiences of war and encourages reflexivity?

## Concepts and narratives

A second contribution that this work has offered is a new method for the analysis of concepts, symbols, and narratives. Conceived as an enquiry into (a) the meanings and (b) the politics of an idea, namely military victory, this book has provided a method to grasp a concept both *in* and *through* time—mobilizing time not as a natural condition in the background of social life, but the product of human-led operations of synthesis that participate in the construction of the social world (see Hom, 2018a, 2020). Unlike approaches that share a similar focus on temporalities (see, for instance, Koselleck-inspired approaches in Berenskoetter, 2016, 2017), however, this method also opened up concept analysis to extratextual sources in general, and material signifiers of victory and traces of time's passage in particular. Indeed, this work explicitly sought to grasp 'the inescapable materiality of war ... [without] relegating it to a second-order effect of [discourse]' (Bousquet, Grove, and Shah, 2020: 102). To that end, following timing theory, I theorized narratives and symbols of victory as both temporal operations of ordering with potent sociopolitical effects, and useful entry points into these timing processes. As a result, this work was able to retrieve and scrutinize victory practices from as varied archival sources as ceremonial parades and artefacts, tattoos, museums, and monuments, doctrine field manuals and history textbooks, commemoration practices and videogames. It did so purposefully, to contextually show that victory travels wartime and peacetime, as well as battlefields and peace zones, therefore leaving the selection of illustrative vignettes to common, and yet varied, empirical manifestation of victory utterances—such as biographical narratives, military strategy, scholarly analyses of the notion, or history teaching materials. This method's remit and applicability, however, is arguably broader—both in terms of furthering research on victory and applying it to other notions.

As far as victory research is concerned, for example, future investigations could broaden the scope of sources under analysis even further. For instance, victory references also abound in movies, TV shows, podcasts, and other products of the entertainment industry, as well as social media content—all relatively understudied materials that would constitute another precious research archive to further the investigation into military victory. Moreover, this method could also deepen the study of conflict-specific victories, perhaps leading to even more varied sources for any specific event under consideration. In turn, this could pave the way to both synchronic and diachronic comparative studies shedding light on country-, area-, culture-, or period-specific trends. Complementarily, defeat, surrender, stalemate, conquest, and other cognate ideas could also be subjected to scrutiny. What is the link between time and these ideas? Does defeat *time* just like victory? In fact, while designed ad hoc for the study of victory—in line with a martial empiricist ethos (Bousquet, Grove, and Shah, 2020)—the method devised here can be also applied to other unrelated concepts. Time is notoriously hyperpervasive, and the list of sociopolitical ideas and categories that, like victory, are tightly related to temporality is likely to be long.

Most importantly, for several scholars across disciplines and traditions, concepts and time are intrinsically connected. Reinhart Koselleck's 'conceptual history', for example, is famously predicated on this link (see, for instance, Koselleck, 2002, 2004). For him, concepts derive their ability to guide action and critique from the double temporal orientation that they have—i.e., the fact that they combine experience and expectation (Koselleck and Richter, 2011: 9). According to Felix Berenskoetter (2017: 154), concepts explicitly 'generat[e] knowledge about the world by organising, naming, and giving meaning to its features'. As Piki Ish-Shalom (2021: 3) put it, concepts are representational and have the specific function of 'mediating between us as sentient beings and the world': they bind 'phenomena into categories that we can understand and process and that we can form expectations' about. On both these accounts, in short, concepts fulfil a function that comes very close to that of timing. In this context, then, this book has also illustrated at least another extrication of this relationship between time and concepts, showing how narratives that underpin 'the conceptions of concepts' (Ish-Shalom, 2021: 1) contribute to 'the sociopolitical construction of the sociopolitical world' (Ish-Shalom, 2021: 2). Which other insights come to light if we put the method that I offered here in conversation with an emerging literature in IR (e.g., Kessler, 2021; Wilkens and Kessler, 2021; see also Berenskoetter, 2016, 2017; Heiskanen, 2021; Ish-Shalom, 2021) interested in concept analysis, from both a methodological and theoretical standpoint?

## Time and temporality

Beyond offering a new ontology of victory and a method to grasp the making of victory *in* and *through* time, this book has also sought to advance the study of time and temporality. In particular, it furthered Hom's (2018a, 2020) timing theory in at least two ways. Firstly, the method discussed above moved timing theory beyond its original focus on textual clues and signifiers of time, with the aim of concomitantly accounting for time's material effects— or 'the material life of time'.[1] Secondly, Chapter 4 introduced a theoretical distinction between *chronic* and *chronological* timing outcomes that admits the possibility that time stands still and does not necessarily flow. Combined, these interventions shed light, for instance, on the notion of wartime and the political mechanisms set in motion through it. I showed that material temporal clues such as military tattoos provide access to war experiences that are chronic, and that, via a wartime hegemonic discourse, they get reinscribed within chronological renderings of war that do not do them justice. Similarly, I clarified how practices of 'temporal othering' (see Hom and Steele, 2016) can be embedded in the design of monuments and other artefacts, for instance, through the politics of construction materials (Chapter 5).

In addition to advancing the (critical) study of war, in other words, this work also furthered the (critical) study of time, yielding new insights into the politics of time and temporality. In so doing, this book also contributed to a burgeoning 'temporal turn' (Berenskoetter, 2011: 664; Hom, 2018b: 305) in IR. In this case too, my theoretical interventions tried to both address ongoing questions and issues that time scholars have recently brought into focus, as well as raise new ones. Showing that multiple wartimes exist, for example, this study has answered Kimberly Hutchings' (2008; see also 2007, 2018) call for a heterotemporal understanding of world politics, offering complementary empirical illustrations of temporal pluralism. Here, this book has also contributed to ongoing debates on the meaning and politics of wartime (see, for instance, Pong, 2020; Rynning, Schmitt, and Theussen, 2021; Hom and Campbell, 2022c; see also Dudziak, 2012; Kaplan, 2018) and the issue of war endings (see, for instance, De Franco, Engberg-Pedersen, and Mennecke, 2019b; Kingsbury and Iron, 2022). Similarly, by centring time in its approach to military victory, this work also contributed to a nascent academic literature that seeks to foreground time and temporality in the study of war and international security (McIntosh, 2016, 2020a; see also Bide, 2021; Caddick, 2021; Jarvis and Robinson, 2021).

Based on these interventions, some avenues for future research can also be identified. Admittedly, for example, much work remains to be done to fully tease out and theorize the various relationships that link narrativity and materiality, especially when it comes to time. As noted early on in this work, for example, queer-theoretical scholarship interested in time and temporality has repeatedly pointed out that time has strong bodily

[1] This phrase was coined by the Temporal Belongings Network—an interdisciplinary research group for the study of time and temporality—as the title of an online conference held in March 2021. More information on the idea of 'the material life of time' and the conference is available at https://www.temporalbelongings.org/the-material-life-of-time.html.

effects (see Rao, 2020), to the point that institutional forms of temporal regulation—or temporalization—of human bodies can be identified throughout history (see Freeman, 2010, 2019). Military institutions are likely to be highly implicated in the administration of comparable temporal regulation practices—through, for example, rigidly timetabled training regimes—that, to date, remain importantly undertheorized (see McNeill, 1995 for an exception). How does war affect soldiers' and other militarized bodies' experience of time? Which forms of embodied timekeeping, beyond military tattoos, can be identified to study these dynamics? Zooming out as to encompass non-martial realms and relations, what are other ways in which time shapes the material world?

Complementarily, new theoretical interventions at the intersection between Quantum Physics and IR have recently started to examine time and temporality more closely, focusing in particular on the 'quantum entanglements' that bind time- and (material) world-making (McIntosh, 2020b; McIntosh, 2022b; see also Brandimarte, 2022 on quantum measurement and processes of identity constitution). This book has offered some illustrations of how, via victory, collective subjects are able to shape the temporal flow they inhabit. While this is certainly one way in which time writing and world-making interlace and overlap, more research should be conducted to examine the relationship between this entanglement. How exactly does time partake in world-making? What are other ways in which—via time—individuals and other social subjects are able to 'shape the world'? Finally, recent interventions on 'discursive martiality' (Millar, 2022) and 'martial politics' (Howell, 2018) have challenged fundamental ideas like militarism and militarization precisely by looking at their temporal politics. This book has showed that, in the case of victory, these politics play out, among other things, via victory practices that reify, mould, and challenge the temporal contours of war. Research on other martial practices that, alongside these, shape and blur war, peace, and other times of world politics remains urgent.

# Bibliography


*11-11 Memories Retold All Endings* (2018). Available at: https://www.youtube.com/watch?v=xSlf-idxV2E (Accessed: 16 November 2022).

British Army (2002). 'Army Field Manual, Volume 1, Part 1, Formation Tactics' Commander Force Development and Training, Ministry of Defence. Available at Liddle Hart Centre for Military Archives, King's College London.

*11-11 Memories Retold Gameplay* (2018). Available at: https://www.youtube.com/watch?v=kL8G-FAUElY (Accessed: 16 November 2022).

Agathangelou, A. M., and Killian, K. D. (eds) (2016). *Time, Temporality and Violence in International Relations: (De)fatalizing the Present, Forging Radical Alternatives.* London: Routledge.

Alexander, C., and Weekes-Bernard, D. (2017). 'History Lessons: Inequality, Diversity and the National Curriculum', *Race Ethnicity and Education*, 20(4): 478–494.

*All Call of Duty Endings* (2022). Available at: https://www.youtube.com/watch?v=gunCD7eeh8g (Accessed: 16 November 2022).

Allison, T. (2010). 'The World War II Video Game, Adaptation, and Postmodern History', *Literature/Film Quarterly*, 38(3): 183–193.

Andrä, C., Bliesemann de Guevara, B., Cole, L., and House, D. (2020). 'Knowing Through Needlework: Curating the Difficult Knowledge of Conflict Textiles', *Critical Military Studies*, 6(3–4): 341–359.

Andrews, M. (2015). 'Poppies, Tommies and Remembrance', *Soundings*, 58: 98–109.

Angelucci, M., and Kerschbamer, S. (2017). 'One Monument, One Town, Two Ideologies: The Monument to the Victory of Bolzano-Bozen', *Public History Review*, 24: 54–75.

Angstrom, J. (2007). 'Introduction: The Problem of Victory and Defeat in Modern War', in J. Angstrom and I. Duyvesteyn (eds), *Understanding Victory and Defeat in Contemporary War.* Abingdon: Routledge. 1–17.

Angstrom, J., and Duyvesteyn, I. (eds) (2007). *Understanding Victory and Defeat in Contemporary War.* Abingdon: Routledge.

Applebaum, A. (2022). 'It's Time to Prepare for a Ukrainian Victory', *The Atlantic*, 11 September. Available at: https://www.theatlantic.com/ideas/archive/2022/09/ukraine-victory-russia-putin/671405/ (Accessed: 25 January 2023).

AQA (n.d.). *AQA in numbers, AQA Realising potential.* Available at: https://www.aqa.org.uk/about-us/who-we-are/aqa-in-numbers (Accessed: 8 October 2022).

Armstrong, M. L., Murphy, K. P., Sallee, A., and Watson, M. G. (2000). 'Tattooed Army Soldiers: Examining the Incidence, Behavior, and Risk', *Military Medicine*, 165(2): 135–141.

Atkinson, M. (2003). *Tattooed: The Sociogenesis of a Body Art.* Toronto: University of Toronto Press.

Australian War Memorial (2020a). *Ink in the Lines | Australian War Memorial, Ink in the Lines.* Available at: https://www.awm.gov.au/visit/exhibitions/ink-in-the-lines (Accessed: 16 January 2021).

Australian War Memorial (2020b). *Miles | Australian War Memorial, Ink in the Lines.* Available at: https://www.awm.gov.au/visit/exhibitions/ink-in-the-lines/miles (Accessed: 15 January 2021).

Australian War Memorial (2020c). *Adam | Australian War Memorial, Ink in the Lines.* Available at: https://www.awm.gov.au/visit/exhibitions/ink-in-the-lines/adam (Accessed: 15 January 2021).

Australian War Memorial (2020d). *Chris | Australian War Memorial, Ink in the Lines.* Available at: https://www.awm.gov.au/visit/exhibitions/ink-in-the-lines/chris (Accessed: 15 January 2021).

Australian War Memorial (2020e). *Newton | Australian War Memorial, Ink in the Lines.* Available at: https://www.awm.gov.au/visit/exhibitions/ink-in-the-lines/newton (Accessed: 15 January 2021).

Axster, S., Danewid, I., Goldstein, A., Mahmoudi, M., Tansel, C. B., and Wilcox, L. (2021). 'Colonial Lives of the Carceral Archipelago: Rethinking the Neoliberal Security State', *International Political Sociology*, 15(3): 415–439.

Balvay, A. (2008). 'Tattooing and Its Role in French: Native American Relations in the Eighteenth Century', *French Colonial History*, 9(1): 1–14.

Bandai Namco (2018). *11-11: MEMORIES RETOLD, Bandai Namco Europe.* Available at: https://en.bandainamcoent.eu/11-11-memories-retold/11-11-memories-retold (Accessed: 16 November 2022).

Barkawi, T. (2011). 'From War to Security: Security Studies, the Wider Agenda and the Fate of the Study of War', *Millennium: Journal of International Studies*, 39(3): 701–716.

Barkawi, T. (2016). 'Decolonising War', *European Journal of International Security*, 1(2): 199–214.

Barkawi, T., and Brighton, S. (2011). 'Powers of War: Fighting, Knowledge, and Critique', *International Political Sociology*, 5(2): 126–143.

Bartholomees, J. B. (2008). 'Theory of Victory', *Parameters*, 38(2): 25–36.

Basham, V. M. (2013). *War, Identity and the Liberal State: Everyday Experiences of the Geopolitical in the Armed Forces.* London: Routledge.

Basham, V.M. (2016). 'Gender, Race, Militarism and Remembrance: The Everyday Geopolitics of the Poppy', *Gender, Place & Culture*, 23(6): 883–896.

Bastian, M., Baraitser, L., Flexer, M. J., Hom, A. R., and Sailsbury, L. (2020). 'Introduction: The Social Life of Time', *Time & Society*, 29(2): 289–296.

BBC (n.d.). *VE Day Broadcasts.* Available at: https://www.bbc.com/historyofthebbc/anniversaries/may/ve-day-broadcasts (Accessed: 15 January 2021).

Beard, M. (2007). *The Roman Triumph.* Cambridge: Belknap Press of Harvard University Press.

Bell, D. S. A. (2003). 'Mythscapes: Memory, Mythology, and National Identity', *The British Journal of Sociology*, 54(1): 63–81.

Bellamy, A. J. (2008). 'The Responsibilities of Victory: *Jus Post Bellum* and the Just War', *Review of International Studies*, 34(4): 601–625.

Berenskoetter, F. (2011). 'Reclaiming the Vision Thing: Constructivists as Students of the Future', *International Studies Quarterly*, 55(3): 647–668.

Berenskoetter, F. (2014). 'Parameters of a National Biography', *European Journal of International Relations*, 20(1): 262–288.

Berenskoetter, F. (ed.) (2016). *Concepts in World Politics.* London: Sage Publications Ltd.

Berenskoetter, F. (2017). 'Approaches to Concept Analysis', *Millennium: Journal of International Studies*, 45(2): 151–173.

Biddle, S. D. (2004). *Military Power: Explaining Victory and Defeat in Modern Battle.* Princeton: Princeton University Press.

Bide, B. (2021). 'In Their Shoes: Using Fashion Objects to Explore the Duration and Complexity of Wartime Experiences', *Critical Military Studies*, 7(4): 418–434.

Bishara, M. (2021). 'The US, the Taliban and the stunning defeat in Afghanistan', *Aljazeera*, 16 August. Available at: https://www.aljazeera.com/opinions/2021/8/16/the-us-the-taliban-and-the-stunning-defeat-in (Accessed: 25 January 2023).

Bloom, W. (1990). *Personal Identity, National Identity and International Relations*. Cambridge: Cambridge University Press.

Blue, S. (2019). 'Institutional Rhythms: Combining Practice Theory and Rhythmanalysis to Conceptualise Processes of Institutionalisation', *Time & Society*, 28(3): 922–950.

Blum, G. (2013). 'The Fog of Victory', *European Journal of International Law*, 24(1): 391–421.

*Bombs to Butterflies* (2014). 'War Ink—"11 for 11"'. Available at: https://vimeopro.com/murga/war-ink-project/page/1 (Accessed: 16 February 2023).

Bond, B. (1996). *The Pursuit of Victory: From Napoleon to Saddam Hussein*. Oxford: Oxford University Press.

Bos, D. (2018). 'Answering the Call of Duty: Everyday Encounters with the Popular Geopolitics of Military-Themed Videogames', *Political Geography*, 63: 54–64.

Bourke, J. (2017). *War and Art: A Visual History of Modern Conflict*. London: Reaktion Books.

Bousquet, A., Grove, J., and Shah, N. (2020). 'Becoming War: Towards a Martial Empiricism', *Security Dialogue*, 51(2–3): 99–118.

Bowen, L., Bradley, K., Middleton, S., Mackillop, A., and Sheldon, N. (2012). 'History in the UK National Curriculum: A Discussion', *Cultural and Social History*, 9(1): 125–143.

Brandimarte, I. (2022). 'Subjects of Quantum Measurement: Surveillance and Affect in the War on Terror', *International Political Sociology*, 16(3): 1–27.

Brandimarte, I. (2023). 'Breathless War: Martial Bodies, Aerial Experiences and the Atmospheres of Empire', *European Journal of International Relations*, 29(3): 525–552.

Brandon, L. (2007). *Art and War*. London: Tauris.

Braunberger, C. (2000). 'Revolting Bodies: The Monster Beauty of Tattooed Women', *NWSA Journal*, 12(2): 1–23.

British Army (2011). 'Army Doctrine Publication AC71954 Army Doctrine Primer'. Commander Force Development and Training, Ministry of Defence.

British Army (2017). 'Army Doctrine Publication AC71940 Land Operations'. Land Warfare Development Centre, Ministry of Defence.

Brown, C. (2017). 'Revisionist Just War Theory and the Impossibility of a Moral Victory', in A. R. Hom, C. O'Driscoll, and K. Mills (eds), *Moral Victories: The Ethics of Winning Wars*. Oxford: Oxford University Press. 85–101.

Bruce, L., and Wilkes, A. (2017). *Oxford AQA GCSE History: Conflict and Tension in Asia 1950–1975*. Oxford: Oxford University Press.

Brüggemann, K., and Kasekamp, A. (2008). 'The Politics of History and the "War of Monuments" in Estonia', *Nationalities Papers*, 36(3): 425–448.

Brüggemann, K., and Kasekamp, A. (2009). 'Identity Politics and Contested Histories in Divided Societies: The Case of Estonian War Monuments', in E. Berg and P. Ehin (eds), *Identity and Foreign Policy: Baltic-Russian Relations and European Integration*. Abingdon: Routledge. 51–63.

Brunstetter, D. R. (2017). 'Justice After the Use of Limited Force: Victory and the Moral Dilemmas of Jus Post Vim', in A. R. Hom, C. O'Driscoll, and K. Mills (eds), *Moral Victories: The Ethics of Winning Wars*. Oxford: Oxford University Press. 214–230.

Budabin, A. C. (2023). 'Heritage Politics in the Case of Black Lives Matter in Bolzano-Bozen, Italy', *European Journal of Cultural and Political Sociology*, 10(4): 577–602.

Bulmer, S., and Jackson, D. (2016). '"You do not live in my skin": Embodiment, Voice, and the Veteran', *Critical Military Studies*, 2(1–2): 25–40.

Burch, S., and Smith, D. J. (2007). 'Empty Spaces and the Value of Symbols: Estonia's "War of Monuments" from Another Angle', *Europe-Asia Studies*, 59(6): 913–936.

BZ '18-'45 (n.d.). 'BZ '18–'45. One Monument, One City, Two Dictatorship'. Available at: https://www.monumenttovictory.com/en.html (Accessed: 22 January 2025).

Caddick, N. (2021). 'Life, Embodiment, and (Post-)War Stories: Studying Narrative in Critical Military Studies', *Critical Military Studies*, 7(2): 155–172.

*Call Of Duty 1 Gameplay* (2016). Available at: https://www.youtube.com/watch?v=PrvX0sanW_E (Accessed: 16 November 2022).

*Call Of Duty 2 Gameplay* (2016). Available at: https://www.youtube.com/watch?v=C97sdjz5b9k (Accessed: 16 November 2022).

*Call of Duty 3 Gameplay* (2020). Available at: https://www.youtube.com/watch?v=oHJyQFY1Llg (Accessed: 16 November 2022).

*Call of Duty Vanguard Gameplay* (2021). Available at: https://www.youtube.com/watch?v=wOhz-U25gvg (Accessed: 16 November 2022).

*Call of Duty WW2 Gameplay* (2017). Available at: https://www.youtube.com/watch?v=X99Rq2wIm4o (Accessed: 16 November 2022).

Campbell, L., and Steele, B. J. (2017). 'The Scars of Victory', in A. R. Hom, C. O'Driscoll, and K. Mills (eds), *Moral Victories: The Ethics of Winning Wars*. Oxford: Oxford University Press. 141–156.

Campbell, L., and Steele, B. J. (2021). 'The Concept of Success in (and of) War', in P. Ish-Shalom (ed.), *Concepts at Work: On the Linguistic Infrastructure of World Politics*. Ann Arbor: University of Michigan Press. 43–64.

Cannadine, D. (2013). 'Making History', *The Times Literary Supplement*, 15 March. 14–15.

Caplan, J. (ed.) (2000). *Written on the Body: The Tattoo in European and American History*. Princeton: Princeton University Press.

Carlà, A., and Mitterhofer, J. (2017). 'Transforming a Controversial Heritage: The Case of the Fascist Victory Monument in South Tyrol', *AUC Studia Territorialia*, 17(2): 11–34.

Caso, F. (2020a). 'Representing Indigenous Soldiers at the Australian War Memorial: A Political Analysis of the Art Exhibition *For Country, For Nation*', *Australian Journal of Political Science*, 55(4): 345–361.

Caso, F. (2020b). 'The Political Aesthetics of the Body of the Soldier', in C. Baker (ed.), *Making War on Bodies: Militarisation, Aesthetics and Embodiment in International Politics*. Edinburgh: Edinburgh University Press. 54–73.

Chakrabarty, D. (2008). *Provincializing Europe: Postcolonial Thought and Historical Difference*. Reissue, with a new preface by the author. Princeton: Princeton University Press.

Chamayou, G. (2015). *Drone Theory*. London: Penguin Books UK.

Chamon, P. (2018). 'Turning Temporal: a Discourse of Time in IR', *Millennium: Journal of International Studies*, 46(3): 396–420.

Choi, A. (2004). 'Democratic Synergy and Victory in War, 1816–1992', *International Studies Quarterly*, 48(3): 663–682.

Cockburn, C. (2004). 'The Continuum of Violence: A Gender Perspective on War and Peace', in W. Giles, and J. Hyndman (eds), *Sites of Violence: Gender and Conflict Zones*. Oakland: University of California Press. 24–44.

Cohen, E. F. (2018). *The Political Value of Time: Citizenship, Duration, and Democratic Justice*. Cambridge: Cambridge University Press.

Coll, S. (2021). 'The Lessons of Defeat in Afghanistan', *The New Yorker*, 5 September. Available at: https://www.newyorker.com/magazine/2021/09/13/the-lessons-of-defeat-in-afghanistan (Accessed: 25 January 2023).

*Commemoration—Ink in the Lines* (2020). Available at: https://www.youtube.com/watch?v=agj2nCIw3jU (Accessed: 16 January 2021).

Connelly, M. (2002). *The Great War Memory and Ritual: Commemoration in the City and East London 1916–1939*. Suffolk: The Royal Historical Society The Boydell Press.

Contested Histories (2021). 'Case Study #87, Victory Monument, Bolzano, Italy'. Available at: https://contestedhistories.org/resources/case-studies/victory-monument-in-bolzano/ (Accessed: 22 January 2025).

Cooper, J. M., and MacMillan, M. (2009). 'Ending the Great War: The Peace That Failed?', in J. Winter (ed.), *The Legacy of the Great War: Ninety Years On*. Columbia, London: University of Missouri Press. 123–158.

Crawford, K. (1995). 'A History of the Right: The Battle for Control of National Curriculum History 1989–1994', *British Journal of Educational Studies*, 43(4): 433–456.

van Creveld, M. L. (1991). *The Transformation of War*. 1st Edition. New York and Toronto: The Free Press.

Cunningham, D. E., Skrede Gleditsch, K., and Salehyan, I. (2009). 'It Takes Two: A Dyadic Analysis of Civil War Duration and Outcome', *Journal of Conflict Resolution*, 53(4): 570–597.

Cuomo, C. J. (1996). 'War Is Not Just an Event: Reflections on the Significance of Everyday Violence', *Hypatia*, 11(4): 30–45.

Daddow, O. J. (2002). 'Facing the Future: History in the Writing of British Military Doctrine', *Defence Studies*, 2(1): 157–164.

Daddow, O. J. (2003). 'British Military Doctrine in the 1980s and 1990s', *Defence Studies*, 3(3): 103–113.

Darda, J. (2019). *Empire of Defense: Race and the Cultural Politics of Permanent War*. Chicago: University of Chicago Press.

*Dare Mighty Things* (2014). 'War Ink—"11 for 11"'. Available at: https://vimeopro.com/murga/war-ink-project/page/1 (Accessed: 16 February 2023).

Daughtry, J.M. (2015). *Listening to War: Sound, Music, Trauma and Survival in Wartime Iraq*. New York: Oxford University Press.

DCDC (2011). 'Joint Doctrine Publication 2-00 Understanding and Intelligence Support to Joint Operations (3rd Edition)'. Development, Concepts and Doctrine Centre, Ministry of Defence. Available at: https://www.gov.uk/government/publications/jdp-2-00-understanding-and-intelligence-support-to-joint-operations (Accessed: 12 March 2022).

DCDC (2013). 'Developing Joint Doctrine Handbook (4th Edition)'. Development, Concepts and Doctrine Centre, Ministry of Defence. Available at: https://www.gov.uk/government/publications/developing-joint-doctrine-handbook (Accessed: 12 March 2022).

DCDC (2014a). 'Joint Doctrine Publication 0-01 UK Defence Doctrine (5th Edition)'. Development, Concepts and Doctrine Centre, Ministry of Defence. Available at: https://www.gov.uk/government/publications/jdp-0-01-fourth-edition-british-defence-doctrine (Accessed: 12 March 2022).

DCDC (2014b). 'Joint Doctrine Publication 01 UK Joint Operations Doctrine'. Development, Concepts and Doctrine Centre, Ministry of Defence. Available at: https://www.gov.uk/government/publications/campaigning-a-joint-doctrine-publication (Accessed: 12 March 2022).

DCDC (2016a). 'Joint Doctrine Publication 04 Understanding and Decision-Making (2nd Edition)'. Development, Concepts and Doctrine Centre, Ministry of Defence. Available at: https://www.gov.uk/government/publications/jdp-04-understanding (Accessed: 12 March 2022).

DCDC (2016b). 'Joint Doctrine Publication 05 Shaping a Stable World: The Military Contribution'. Development, Concepts and Doctrine Centre, Ministry of Defence. Available at: https://www.gov.uk/government/publications/shaping-a-stable-world-the-military-contribution-jdp-05 (Accessed: 12 March 2022).

DCDC (2017a). 'Joint Doctrine Publication 0-10 UK Maritime Power'. Development, Concepts and Doctrine Centre, Ministry of Defence. Available at: https://www.gov.uk/government/publications/uk-maritime-power-jdp-0-10 (Accessed: 12 March 2022).

DCDC (2017b). 'Joint Doctrine Publication 0-20 UK Land Power'. Development, Concepts and Doctrine Centre, Ministry of Defence. Available at: https://www.gov.uk/government/publications/uk-land-power-jdp-0-20 (Accessed: 12 March 2022).

DCDC (2017c). 'Joint Doctrine Publication 0-30 UK Air and Space Power'. Development, Concepts and Doctrine Centre, Ministry of Defence. Available at: https://www.gov.uk/government/publications/uk-air-and-space-doctrine-jdp-0-30 (Accessed: 12 March 2022).

DCDC (2018). 'Joint Doctrine Publication 3-46 Legal Support to Joint Operations (3rd Edition)'. Development, Concepts and Doctrine Centre, Ministry of Defence. Available at: https://www.gov.uk/government/publications/jdp-3-46-legal-support-to-joint-operations-2nd-edition (Accessed: 12 March 2022).

DCDC (2019). 'Joint Doctrine Note 2/19 Defence Strategic Communication: an Approach to Formulating and Executing Strategy'. Development, Concepts and Doctrine Centre, Ministry of Defence. Available at: https://www.gov.uk/government/publications/jdn-219-defence-strategic-communication-an-approach-to-formulating-and-executing-strategy (Accessed: 12 March 2022).

DCDC (2021a). 'Integrated Operating Concept'. Development, Concepts and Doctrine Centre, Ministry of Defence. Available at: https://www.gov.uk/government/publications/the-integrated-operating-concept-2025 (Accessed: 12 March 2022).

DCDC (2021b). 'Joint Doctrine Publication 02 UK Operations: The Defence Contribution to Resilience (4th Edition)'. Development, Concepts and Doctrine Centre, Ministry of Defence. Available at: https://www.gov.uk/government/publications/operations-in-the-uk-a-joint-doctrine-publication (Accessed: 12 March 2022).

De Franco, C., Engberg-Pedersen, A., and Mennecke, M. (2019a). 'How Do Wars End? A Multidisciplinary Enquiry', *Journal of Strategic Studies*, 42(7): 889–900.

De Franco, C., Engberg-Pedersen, A., and Mennecke, M. (eds) (2019b). 'Special Issue: How Do Wars End? A Multidisciplinary Enquiry', *Journal of Strategic Studies*, 42(7): 889–1037.

DeMello, M. (2000). *Bodies of Inscription: A Cultural History of the Modern Tattoo Community*. Durham: Duke University Press.

Department for Education (2014a). 'GCE AS and A level subject content for history'. Available at: https://www.gov.uk/government/publications/gce-as-and-a-level-for-history (Accessed: 16 October 2022).

Department for Education (2014b). 'History GCSE subject content'. Available at: https://www.gov.uk/government/publications/gcse-history (Accessed: 16 October 2022).

Department for Education (2014c). 'The national curriculum in England: Key stages 3 and 4 framework document'. Available at: https://www.gov.uk/government/publications/national-curriculum-in-england-secondary-curriculum (Accessed: 16 October 2022).

Desch, M. C. (2002). 'Democracy and Victory: Why Regime Type Hardly Matters', *International Security*, 27(2): 5–47.

Desch, M. C. (2003). 'Democracy and Victory: Fair Fights or Food Fights?', *International Security*, 28(1): 180–194.

Dierkes, J. (2010). *Postwar History Education in Japan and the Germanys: Guilty Lessons*. Abingdon: Routledge.

Di Michele, A. (2020). 'Storicizzare i monumenti fascisti: Il caso di Bolzano', *Storia e Regione*, 29(2): 149–167.

*Doc* (2014). 'War Ink—"11 for 11"'. Available at: https://vimeopro.com/murga/war-ink-project/page/2 (Accessed: 16 February 2023).

Downes, A. B. (2009). 'How Smart and Tough Are Democracies? Reassessing Theories of Democratic Victory in War', *International Security*, 33(4): 9–51.

Dudziak, M. L. (2012). *War Time: An Idea, Its History, Its Consequences*. Oxford: Oxford University Press.

Dunn, R. E. (2000). 'The Making of a National Curriculum: The British Case', *The History Teacher*, 33(3): 395–398.

Dyvik, S. L. (2016). '"Valhalla Rising": Gender, Embodiment and Experience in Military Memoirs', *Security Dialogue*, 47(2): 133–150.

Dyvik, S. L., and Welland, J. (2018). 'War Ink: Sense-Making and Curating War Through Military Tattoos', *International Political Sociology*, 12(4): 346–361.

Eckert, A. E. (2017). 'Cui Bono: Moral Victory in Privatized War', in A. R. Hom, C. O'Driscoll, and K. Mills (eds), *Moral Victories: The Ethics of Winning Wars*. Oxford: Oxford University Press. 198–214.

Edge Staff (2015). *The Making Of... Medal of Honor, gamesradar*. Available at: https://www.gamesradar.com/making-medal-honor/ (Accessed: 16 November 2022).

Edkins, J. (2003). *Trauma and the Memory of Politics*. Cambridge: Cambridge University Press.

Ehala, M. (2009). 'The Bronze Soldier: Identity Threat and Maintenance in Estonia', *Journal of Baltic Studies*, 40(1): 139–158.

Electronic Arts (n.d.). *Medal of Honor Video Games—Official EA Site Electronic Arts Inc.*, Available at: https://www.ea.com/en-gb/games/medal-of-honor (Accessed: 16 November 2022).

Elias, N. (1978). *What Is Sociology?* New York: Columbia University Press.

Elias, N. (2007). *An Essay on Time*. Dublin: University College Dublin Press.

El-Shazly, N. E. S. (2016). *The Gulf Tanker War: Iran and Iraq's Maritime Swordplay*. London: Palgrave Macmillan.

Engelhardt, T. (2007). *The End of Victory Culture: Cold War America and the Disillusioning of a Generation*. 2nd Edition. Amherst: University of Massachusetts Press.

England, J. (2018). *An Interview with the Creators Of 11-11: Memories Retold | New Rising Media*. Available at: https://www.newrisingmedia.com/blog/2018/12/1/an-interview-with-the-creators-of-11-11-memories-retold (Accessed: 16 November 2022).

Fanon, F., and Philcox, R. (2004). *The Wretched of the Earth: Frantz Fanon*. Translated from the French by Richard Philcox; introductions by Jean-Paul Sartre and Homi K. Bhabha. New York: Grove Press.

Farrell, T. (2008). 'The dynamics of British military transformation', *International Affairs*, 84(4): 777–807.

Favargiotti, S., Busana, A., and Cappelletti, D. (2020). 'Beyond Italianization: Conflicts, Stories, and Reactions of the Afterlives of Fascism in Bolzano/Bozen', in K. B. Jones and S. Pilat (eds), *The Routledge Companion to Italian Fascist Architecture*. Abingdon: Routledge. 198–211.

Ferriby, D., Martin, D., and Walsh, B. (2016). *AQA GCSE History: Understanding the Modern World*. UK Edition. London: Hodder Education.

Filkins, D. (2008). *The Forever War*. 1st Edition. New York: Alfred a Knopf Inc.

Fletcher, J.R., Zubair, M., and Roche, M. (2022). 'The Neuropsychiatric Biopolitics of Dementia and Its Ethnicity Problem', *The Sociological Review*, 70(5): 1005–1024.

Ford, M., and Hoskins, A. (2022). *Radical War: Data, Attention and Control in the Twenty-First Century*. London: C Hurst & Co Publishers Ltd.

Forest, B., and Johnson, J. (2002). 'Unraveling the Threads of History: Soviet–Era Monuments and Post–Soviet National Identity in Moscow', *Annals of the Association of American Geographers*, 92(3): 524–547.

Freeman, E. (2010). *Time Binds: Queer Temporalities, Queer Histories*. Durham: Duke University Press.
Freeman, E. (2019). *Beside you in time: sense methods and queer sociabilities in the American nineteenth century*. Durham: Duke University Press.
Gadd, M. C. (1992). 'A Survey of Soldiers' Attitudes to Tattooing', *Journal of the Royal Army Medical Corps*, 138(2): 73–76.
Gagnon, C., Senders, S., and Brown, K. (2014), 'Introduction' in C. Gagnon and K. Brown (eds), *Post-Conflict Studies: An Interdisciplinary Approach*. London and New York: Routledge. 1–16.
Galai, Y. (2019). 'The victory Image: Imaging Israeli Warfighting from Lebanon to Gaza', *Security Dialogue*, 50(4): 295–313.
Galai, Y. (2021). 'On "Victory Images" and Gaza', *Journal of Visual Culture*, 20(2): 211–215.
Gell, A. (1993). *Wrapping in Images: Tattooing in Polynesia*. Oxford, New York: Clarendon Press.
Gent, S. E. (2008). 'Going in When It Counts: Military Intervention and the Outcome of Civil Conflicts', *International Studies Quarterly*, 52(4): 713–735.
Gerwarth, R. (2016). *The Vanquished: Why the First World War Failed to End*. New York: Farrar, Straus & Giroux.
Gerwarth, R., and Horne, J. (2012). *War in Peace: Paramilitary Violence in Europe after the Great War*. Oxford: Oxford University Press.
Gibler, D. M., and Miller, S. V. (2013). 'Quick Victories?: Territory, Democracies, and Their Disputes', *Journal of Conflict Resolution*, 57(2): 258–284.
Gibson-Fall, F. (2021). 'Military Responses to COVID-19, Emerging Trends in Global Civil-Military Engagements', *Review of International Studies*, 47(2): 155–170.
Gish, H. (2010). 'Playing the Second World War: Call of Duty and the Telling of History', *Eludamos: Journal for Computer Game Culture*, 4(2): 167–180.
Glennie, P., and Thrift, N. J. (2009). *Shaping the Day: A History of Timekeeping in England and Wales 1300–1800*. Oxford, New York: Oxford University Press.
Goncharova, O. (2022). 'Zelensky to US Congress: "Your money is not charity, it's investment in global security"', *The Kyiv Independent*, 22 December. Available at: https://kyivindependent.com/national/zelensky-to-us-congress (Accessed: 15 January 2023).
Gooch, J. (1997). 'Introduction: Military Doctrine and Military History', in J. Gooch (ed.), *Strategic and Combat Studies Institute Occasional Paper Number 30: The Origins of Contemporary Doctrine*. UK Strategic and Combat Studies Institute. 5–6.
Govenar, A.B. (1982). 'The Changing Image of Tattooing in American Culture', *The Journal of American Culture*, 5(1): 30–37.
Gray, C. (2002). *Defining and Achieving Decisive Victory*. Fort Belvoir, VA: Defense Technical Information Center.
Gray, H. (2019). 'The "War"/"Not-War" Divide: Domestic Violence in the Preventing Sexual Violence Initiative', *The British Journal of Politics and International Relations*, 21(1): 189–206.
Greenberg, A. (1989). 'Lutyens's Cenotaph', *Journal of the Society of Architectural Historians*, 48(1): 5–23.
Gregory, A. (1994). *The Silence of Memory: Armistice Day 1919–1946*. Oxford, Providence: Berg Publishers.
Gregory, T. (2020). 'The Costs of War: Condolence Payments and the Politics of Killing Civilians ', *Review of International Studies*, 46(1): 156–176.
Gregory, T. (2022). 'Calibrating Violence: Body Counts as a Weapon of War', *European Journal of International Security*, 7(4): 479–507.

Griffen, W. L. (1974). 'Telling It like It Was: Vietnam History in the Schools', *The Phi Delta Kappan*, 55(9): 616–618.

gruppe gut (2014). 'Progetto di Allestimento per un Percorso Espositivo nel Monumento alla Vittoria di Bolzano'. Available at: https://www.monumentoallavittoria.com/fileadmin/user_upload/pdfs/GGG_MaV_Concetto_ITA.2.pdf (Accessed: 22 January 2025).

The Guardian (2022). 'The Guardian view on the Taliban's victory: Afghanistan's nightmare is intensifying', The Guardian, 12 August. Available at: https://www.theguardian.com/commentisfree/2022/aug/12/the-guardian-view-on-the-talibans-victory-afghanistans-nightmare-is-intensifying (Accessed: 25 January 2023).

Guarnieri, A. (2017). 'Il referendum su "piazza della Pace" e la memoria divisa dell'Alto Adige: una *Never-Ending Story*?', *Storie e Regione*, 26(2): 24–32.

Haight, A., Wright, S., Aldridge, D., and Alexander, P. (2021). 'Remembrance Day Practices in Schools: Meaning-Making in Social Memory During the First World War Centenary', *Journal of Beliefs & Values*, 42(1): 33–48.

Hall, J. (2022). '"Winning" the Forever Wars? Presidential Rhetoric and US Ontological (In)Security', *Global Studies Quarterly*, 2(4): 1–14.

Hall, J. (2024). 'The War on Terror and the Victory Trap', *Foreign Policy Analysis*, 20(3): 1–20. Available at: https://doi.org/10.1093/fpa/orae012 (Accessed: 2 April 2025).

Hardman, R. (2013). 'Why we SHOULD upset the Germans—by reminding them of their Great War atrocities', *Daily Mail*, 14 January. Available at: https://www.dailymail.co.uk/news/article-2262443/Why-SHOULD-upset-Germans—reminding-Great-War-atrocities.html (Accessed: 16 October 2022).

Harel, A., and Issacharoff, A. (2008). *34 Days: Israel, Hezbollah, and the War in Lebanon*. 1st Edition. New York: Palgrave Macmillan.

Hartnett, L., Glanville, L., O'Driscoll, C., Wilcox, L., Bellamy, A., and Steele, B. (2022). 'Winning? The Politics of Victory in an Era of Endless War', *International Studies Review*, 24(1): 1–25.

Hasian, M. (2010). 'U.S. Military Perceptions of Victories in Iraq, the "Long War" Against Terrorism, and the Enduring Rhetorical Power of the 2006 Counterinsurgency Manual', *Western Journal of Communication*, 74(5): 570–587.

Hein, L. E., and Selden, M. (eds) (2000). *Censoring History: Citizenship and Memory in Japan, Germany, and the United States*. London: M.E. Sharpe.

Heiskanen, J. (2021). 'Found in Translation: The Global Constitution of the Modern International Order', *International Theory*, 13(2): 231–259.

Hellen, N., and Brooks, R. (2013). 'Don't mention that we won the First World War', *The Sunday Times*, 13 January. Available at: https://www.thetimes.co.uk/article/dont-mention-that-we-won-the-first-world-war-hh3r8vcnz3v (Accessed: 16 October 2022).

Henderson, E. A., and Bayer, R. (2013). 'Wallets, Ballots, or Bullets: Does Wealth, Democracy, or Military Capabilities Determine War Outcomes?, *International Studies Quarterly*, 57(2): 303–317.

Hess, A. (2007). '"You Don't Play, You Volunteer": Narrative Public Memory Construction in Medal of Honor: Rising Sun', *Critical Studies in Media Communication*, 24(4): 339–356.

Heuser, B. (2013). 'Victory, Peace, and Justice: The Neglected Trinity', *Joint Forces Quarterly*, 69: 1–7.

Heuser, B. (2017). 'Defeat as Moral Victory: The Historical Experience', in A. R. Hom, C. O'Driscoll, and K. Mills (eds) *Moral Victories: The Ethics of Winning Wars*. Oxford: Oxford University Press. 52–69.

Hirst, A. (2021). '"Videogames Saved My Life": Everyday Resistance and Ludic Recovery among US Military Veterans', *International Political Sociology*, 15(4): 482–503.

Hobbs, R. W. (1979). *The Myth of Victory: What Is Victory in War?* Boulder: Westview Press.

Hockey, J. (2009). '"Switch On": Sensory Work in the Infantry', *Work, Employment and Society*, 23(3): 477–493.
Höglund, J. (2008). 'Electronic Empire: Orientalism Revisited in the Military Shooter', *Game Studies*, 8(1).
Høiback, H. (2011). 'What Is Doctrine?', *Journal of Strategic Studies*, 34(6): 879–900.
Høiback, H. (2013). *Understanding Military Doctrine: A Multidisciplinary Approach.* London: Routledge.
Hökerberg, H. (2017). 'The Monument to Victory in Bolzano: Desacralisation of a Fascist Relic', *International Journal of Heritage Studies*, 23(8): 759–774.
Hölscher, T. (2006). 'The Transformation of Victory into Power: From Event to Structure', in S. Dillon and K.E. Welch (eds), *Representations of War in Ancient Rome.* Cambridge: Cambridge University Press. 27–48.
Hom, A. R. (2010). 'Hegemonic Metronome: The Ascendancy of Western Standard Time', *Review of International Studies*, 36(4): 1145–1170.
Hom, A. R. (2016). 'Angst Springs Eternal: Dangerous Times and the Dangers of Timing the "Arab Spring"', *Security Dialogue*, 47(2): 165–183.
Hom, A. R. (2017). 'Patriots All Around: Inter/national Timing, Round Numbers, and the Politics of Commemorative Critique', *Australian Journal of Politics & History*, 63(3): 443–456.
Hom, A. R. (2018a). 'Timing Is Everything: Toward a Better Understanding of Time and International Politics', *International Studies Quarterly*, 62(1): 69–79.
Hom, A. R. (2018b). 'Silent Order: The Temporal Turn in Critical International Relations', *Millennium: Journal of International Studies*, 46(3): 303–330.
Hom, A. R. (2020). *International Relations and the Problem of Time.* Oxford, New York: Oxford University Press.
Hom, A. R., and Campbell, L. (2022a). 'Wartime in the 21st Century', *International Relations*, 36(4): 525–546.
Hom, A. R., and Campbell, L. (2022b). 'The Liturgy of Triumph: Victory Culture, Popular Rituals, and the US Way of Wartiming', *International Relations*, 36(4): 591–615.
Hom, A. R., and Campbell, L. (eds) (2022c). 'Special Issue: The Politics of Wartime in the 21st Century', *International Relations*, 36(4): 525–686.
Hom, A. R., McIntosh, C., McKay, A., and Stockdale, L. (eds) (2016). *Time, Temporality and Global Politics.* Bristol: E-International Relations Publishing.
Hom, A. R., O'Driscoll, C., and Mills, K. (eds) (2017). *Moral Victories: The Ethics of Winning Wars.* Oxford: Oxford University Press.
Hom, A. R., and Solomon, T. (2016). 'Timing, Identity, and Emotion in International Relations', in A. R. Hom et al. (eds), *Time, Temporality and Global Politics.* Bristol: E-International Relations Publishing. 20–37.
Hom, A. R., and Steele, B. J. (2016). 'Child's Play: Temporal Discourse, Counterpower, and Environmental Politics', in A. M. Agathangelou and K. D. Killian (eds), *Time, Temporality and Violence in International Relations: (De)Fatalizing the Present, Forging Radical Alternatives.* London: Routledge. 189–204.
Hom, A. R., and Steele, B. J. (2020). 'Anxiety, Time, and Ontological Security's Third-Image Potential', *International Theory*, 12(2): 322–336.
Homberger, E. (1976). 'The Story of the Cenotaph', *The Times Literary Supplement*, 12 November. 1429–1430.
Howard, D. (1919). 'Armistice Day', *The Times*, 4 November. 8.
Howard, M. (1999). 'When Are Wars Decisive?', *Survival*, 41(1): 126–135.
Howell, A. (2018). 'Forget "Militarization": Race, Disability and the "Martial Politics" of the Police and of the University', *International Feminist Journal of Politics*, 20(2): 117–136.

Hukelova, M., and Holloway, M. (2021). '"Lest We Forget": The Significance and Meaning of Remembrance Sunday in Contemporary British Society', *Mortality*. 1–19.

Huntemann, N., and Payne, M. T. (eds) (2010). *Joystick Soldiers: The Politics of Play in Military Video Games*. New York: Routledge.

Hutchings, K. (2007). 'Happy Anniversary! Time and Critique in International Relations Theory', *Review of International Studies*, 33(S1): 71–89.

Hutchings, K. (2008). *Time and World Politics: Thinking the Present*. Manchester: Manchester University Press.

Hutchings, K. (2018). 'Time and the Study of World Politics', *Millennium: Journal of International Studies*, 46(3): 253–258.

Hutchison, E., and Calkivik, A. (eds) (2019). 'Making War, Making Sense? Debating Jens Bartelson's *War in International Thought*', *Millennium: Journal of International Studies*, 48(1): 60–114.

Iasiello, L. V. (2004). 'Jus Post Bellum', *Naval War College Review*, 57(3): 33–52.

*Identity and Belonging—Ink in the Lines* (2020). Available at: https://www.youtube.com/watch?v=Ge9giZ-c6-8 (Accessed: 15 January 2021).

Illinois Library (2016a). *LibGuides: Symbols of Service Exhibit: John*. Available at: https://guides.library.illinois.edu/c.php?g=590394&p=4081865 (Accessed: 15 January 2021).

Illinois Library (2016b). *LibGuides: Symbols of Service Exhibit: Garrett*. Available at: https://guides.library.illinois.edu/c.php?g=590394&p=4081871 (Accessed: 15 January 2021).

Illinois Library (2016c). *LibGuides: Symbols of Service Exhibit: Home*. Available at: https://guides.library.illinois.edu/c.php?g=590394&p=4081863 (Accessed: 14 January 2021).

Illinois Library (2016d). *LibGuides: Symbols of Service Exhibit: Brent*. Available at: https://guides.library.illinois.edu/c.php?g=590394&p=4081869 (Accessed: 14 January 2021).

Illinois Library (2016e). *LibGuides: Symbols of Service Exhibit: Caleb*. Available at: https://guides.library.illinois.edu/c.php?g=590394&p=4081867 (Accessed: 14 January 2021).

Imperial War Museum (2022). 'Exhibition Leaflet—War Games: Real Conflicts, Virtual Worlds, Extreme Entertainment'.

Imperial War Museum (n.d.). *War Games, Imperial War Museums*. Available at: https://www.iwm.org.uk/events/war-games (Accessed: 15 November 2022).

*Ink in the Lines—Exhibition* (2020). Available at: https://www.youtube.com/watch?v=8rVc0rMsrZI (Accessed: 15 January 2021).

International Churchill Society (n.d.). *This Is Your Victory, International Churchill Society*. Available at: https://winstonchurchill.org/resources/speeches/1941-1945-war-leader/this-is-your-victory/ (Accessed: 15 January 2021).

*Interview with creators of War Ink* (2014). Available at: https://vimeo.com/114941114 (Accessed: 14 January 2021).

Ish-Shalom, P. (ed.) (2021). *Concepts at Work: On the Linguistic Infrastructure of World Politics*. Ann Arbor: University of Michigan Press.

Jacques, M. (2021). 'Defeat in Afghanistan a complete humiliation for the US', *Global Times*, 15 August. Available at: https://www.globaltimes.cn/page/202108/1231540.shtml (Accessed: 25 January 2023).

Jarvis, L., and Robinson, N. (2021). 'War, Time, and Military Videogames: Heterogeneities and Critical Potential', *Critical Military Studies*, 7(2): 192–211.

Johnson, D. D. P., and Tierney, D. (2006). *Failing to Win: Perceptions of Victory and Defeat in International Politics*. Cambridge: Harvard University Press.

Johnson, D. D. P., and Tierney, D. (2007). 'In the Eye of the Beholder: Victory and Defeat in Us Military Operations', in J. Angstrom and I. Duyvesteyn (eds), *Understanding Victory and Defeat in Contemporary War*. Abingdon: Routledge. 46–76.

Johnson, J. T. (2017). 'Victory Though the Heavens Fall? Unlimited Warfare as Theme and Phenomenon', in A. R. Hom, C. O'Driscoll, and K. Mills (eds), *Moral Victories: The Ethics of Winning Wars*. Oxford: Oxford University Press. 69–85.

Johnston, P. (2000). 'Doctrine Is Not Enough: The Effect of Doctrine on the Behavior of Armies', *Parameters*, 30(3): 30–39.

Jones, B. T. (2017). 'Altering Capabilities or Imposing Costs? Intervention Strategy and Civil War Outcomes', *International Studies Quarterly*, 61(1): 52–63.

Kang, M., and Jones, K. (2007). 'Why Do People Get Tattoos?', *Contexts*, 6(1): 42–47.

Kaplan, C. (2018). *Aerial Aftermaths: Wartime from Above*. Durham: Duke University Press.

Karsh, E. (2002). *The Iran-Iraq War, 1980–1988*. Oxford: Osprey Pub.

Kattago, S. (2009). 'War Memorials and the Politics of Memory: The Soviet War Memorial in Tallinn', *Constellations*, 16(1): 149–166.

Katz, Y. (2006). 'Analysis: The need to show a victory', *The Jerusalem Post*, 26 July. Available at: https://www.jpost.com/Israel/Analysis-The-need-to-show-a-victory (Accessed: 8 January 2023).

Kearey, K. (2015). *Oxford AQA History A Level and AS Component 2: International Relations and Global Conflict c1890–1941*. Oxford: Oxford University Press.

Kelly, J. (2013). 'Popular Culture, Sport and the 'Hero'-fication of British Militarism', *Sociology*, 47(4): 722–738.

Kelly, J. (2023). 'A Critical Discourse Analysis of Military-Related Remembrance Rhetoric in UK Sport: Communicating Consent for British Militarism', *Communication & Sport*, 11(1): 192–212.

Kempshall, C. (2019). 'Centenary (Computer Games)', *International Encyclopedia of the First World War (Online)*. Berlin: Freie Universität Berlin.

Kessler, O. (2021). 'Conceptual History in International Relations: From Ideology to Social Theory?', in B. de Carvalho, J. Costa Lopez, and H. Leira (eds), *Routledge Handbook of Historical International Relations*. Abingdon: Routledge. 551–561.

Khoury, D. R. (2013). *Iraq in Wartime: Soldiering, Martyrdom, and Remembrance*. Cambridge: Cambridge University Press.

Kingsbury, D., and Iron, R. (eds) (2022). *How Wars End: Theory and Practice*. London: Routledge.

Kirby, P. (2020). 'The Body Weaponized: War, Sexual Violence and the Uncanny', *Security Dialogue*, 51(2–3): 211–230.

Kirkpatrick, T. (2018). 'Here's the origin of the respected battlefield cross', *We Are The Mighty*, 18 June. Available at: https://www.wearethemighty.com/popular/origin-of-the-battlefield-cross/ (Accessed: 17 January 2021).

Kober, A. (2008). 'The Israel Defense Forces in the Second Lebanon War: Why the Poor Performance?', *Journal of Strategic Studies*, 31(1): 3–40.

Kolenda, C. D. (2019). 'Slow Failure: Understanding America's Quagmire in Afghanistan', *Journal of Strategic Studies*, 42(7): 992–1014.

Koselleck, R. (2002). *The Practice of Conceptual History*. Stanford: Stanford University Press.

Koselleck, R. (2004). *Futures past: On the Semantics of Historical Time*. New York: Columbia University Press.

Koselleck, R., and Richter, M. (2011). 'Basic Concepts in History: A Historical Dictionary of Political and Social Language in Germany', *Contributions to the History of Concepts*, 6(1): 1–37.

Kosut, M. (2000). 'Tattoo Narratives: The Intersection of the Body, Self-Identity and Society', *Visual Sociology*, 15(1): 79–100.

Krause, J. (2019). 'How Do Wars End? A Strategic Perspective', *Journal of Strategic Studies*, 42(7): 920–945.

Krause, U. (2015). 'A Continuum of Violence? Linking Sexual and Gender-based Violence during Conflict, Flight, and Encampment', *Refugee Survey Quarterly*, 34(4): 1–19.

Krook, M. L. (2020). *Violence Against Women in Politics*. Oxford: Oxford University Press.

Krystalli, R. C. (2021). 'Narrating Victimhood: Dilemmas and (in)Dignities', *International Feminist Journal of Politics*, 23(1): 125–146.

Krystalli, R. (2024). *Good Victims: The Political as a Feminist Question*. Oxford: Oxford University Press.

Kwon, H. (2008). *Ghosts of War in Vietnam*. Cambridge: Cambridge University Press.

Lake, D. A. (1992). 'Powerful Pacifists: Democratic States and War', *American Political Science Review*, 86: 24–37.

Lake, D. A. (2003). 'Fair Fights? Evaluating Theories of Democracy and Victory', *International Security*, 28(1): 154–167.

Lazar, N. C. (2019). *Out of Joint: Power, Crisis and the Rhetoric of Time*. New Haven: Yale University Press.

Lehti, M., Jutila, M., and Jokisipilä, M. (2008). 'Never-Ending Second World War: Public Performances of National Dignity and the Drama of the Bronze Soldier', *Journal of Baltic Studies*, 39(4): 393–418.

Leisten, T. (1996). 'Mashhad Al-Nasr: Monuments of War and Victory in Medieval Islamic Art', *Muqarnas*, 13: 7–26.

Lemma, A. (2010). *Under the Skin: A Psychoanalytic Study of Body Modification*. London, New York: Routledge.

Leser, J., and Sterrett, J. (2010). 'A Battle in Every Classroom: Gaming and the U.S. Army Command & General Staff College', in N. Huntemann and M. T. Payne (eds), *Joystick Soldiers: The Politics of Play in Military Video Games*. New York: Routledge. 146–159.

Lidher, S., McIntosh, M., and Alexander, C. (2021). 'Our Migration Story: History, the National Curriculum, and Re-Narrating the British Nation', *Journal of Ethnic and Migration Studies*, 47(18): 4221–4237.

Lisle, D. (2016). *Holidays in the Danger Zone: Entanglements of War and Tourism*. Minneapolis, London: University of Minnesota Press.

Loewen, J. W. (2000). 'The Vietnam War in High School American History', in L. E. Hein and M. Selden (eds), *Censoring History: Citizenship and Memory in Japan, Germany, and the United States*. London: M.E. Sharpe. 150–172.

Loewen Walker, R. (2014). The Living Present as a Materialist Feminist Temporality. *Women: A Cultural Review*, 25(1), 46–61.

Longley, E. (2016). *Oxford AQA GCSE History: Conflict and Tension 1918–1939*. Oxford: Oxford University Press.

MacKenzie, M., Gregory, T., Shah, N., Barkawi, T., Haastrup, T., Eichler, M., Wegner, M., and Howell, A. (2019). 'Can We Really "Forget" Militarization? A Conversation on Alison Howell's Martial Politics', *International Feminist Journal of Politics*, 21(5): 816–836.

MacLeish, K. T. (2013). *Making War at Fort Hood: Life and Uncertainty in a Military Community*. Princeton: Princeton University Press.

Makiya, K. (2004). *The Monument: Art and Vulgarity in Saddam Hussein's Iraq*. New Edition. London: Tauris.

Manchanda, N. (2015). 'Queering the Pashtun: Afghan Sexuality in the Homo-Nationalist Imaginary', *Third World Quarterly*, 36(1): 130–146.

Manchanda, N., and Rossdale, C. (2021). 'Resisting Racial Militarism: War, Policing and the Black Panther Party', *Security Dialogue*, 52(6): 473–492.

Mandel, R. (2006). *Meaning of Military Victory*. 1st Edition. Boulder: Lynne Rienner Publishers.

Mandel, R. (2007). 'Reassessing Victory in Warfare', *Armed Forces & Society*, 33(4): 461–495.

Mansell, J. G. (2009). 'Musical Modernity and Contested Commemoration at the Festival of Remembrance, 1923–1927', *The Historical Journal*, 52(2): 433–454.
Marshall, D. (2004). 'Making Sense of Remembrance', *Social & Cultural Geography*, 5(1): 37–54.
Martel, W. C. (2011a). *Victory in War: Foundations of Modern Strategy*. Expanded Edition. Cambridge, New York: Cambridge University Press.
Martel, W. C. (2011b). 'Victory in Scholarship on Strategy and War', *Cambridge Review of International Affairs*, 24(3): 513–536.
*Mateship—Ink in the Lines* (2020). Available at: https://www.youtube.com/watch?v=TDQT8WIkWHY (Accessed: 15 January 2021).
McCormick, M. (1986). *Eternal Victory: Triumphal Rulership in late Antiquity, Byzantium and the early Medieval West*. Cambridge: Cambridge University Press.
McIntosh, C. (2016). 'War Through a Temporal Lens: Foregrounding Temporality in International Relations' Conceptions of War', in A. R. Hom et al. (eds), *Time, Temporality and Global Politics*. Bristol: E-International Relations Publishing. 115–133.
McIntosh, C. (2020a). 'Theorizing the Temporal Exception: The Importance of the Present for the Study of War', *Journal of Global Security Studies*, 5(4): 543–558.
McIntosh, C. (2020b). 'Writing Quantum Entanglement into International Relations: Temporality, Positionality, and the Ontology of War', *Millennium: Journal of International Studies*, 49(1): 162–174.
McIntosh, C. (2022a). 'A "Continuing, Imminent" Threat: The Temporal Frameworks Enabling the US War on Terrorism', *International Relations*, 36(4): 568–590.
McIntosh, C. (2022b). 'Why the Temporal Turn in IR Should Care about Quantum Theory, and Vice Versa', *E-International Relations*, 19 April. Available at: https://www.e-ir.info/2022/04/19/why-the-temporal-turn-in-ir-should-care-about-quantum-theory-and-vice-versa/ (Accessed: 25 January 2023).
McIntosh, C. (2023). *The Time of Global Politics: International Relations as Study of the Present*. Cambridge: Cambridge University Press.
McNeill, W. H. (1995). *Keeping Together in Time: Dance and Drill in Human History*. Cambridge: Harvard University Press.
McSorley, K. (2012). 'Helmetcams, Militarized Sensation and "Somatic War"', *Journal of War & Culture Studies*, 5(1): 47–58.
McSorley, K. (2013). 'War and the Body', in K. McSorley (ed.), *War and the Body: Militarisation, Practice and Experience*. London: Routledge. 1–31.
McSorley, K. (2014). 'Towards an Embodied Sociology of War', *The Sociological Review*, 62(2 suppl): 107–128.
McSorley, K. (2020). 'Sensate Regimes of War: Smell, Tracing and Violence', *Security Dialogue*, 51(2–3): 155–173.
*Medal of Honor Gameplay* (2017). Available at: https://www.youtube.com/watch?v=x6ivwknQN54 (Accessed: 16 November 2022).
Mifflin, M. (2001). *Bodies of Subversion: A Secret History of Women and Tattoo*. 2nd Edition. New York: Umbrage Editions Inc.
Mileham, P. (2004a). 'Fifty Years of British Army Officership 1960–2010', *Defense & Security Analysis*, 20(1): 69–86.
Mileham, P. (2004b). 'Fifty Years of British Army Officership 1960–2010: Part II: Prospective', *Defense & Security Analysis*, 20(2): 179–199.
Millar, K. M. (2022). *Support the Troops: Military Obligation, Gender, and the Making of Political Community*. Oxford: Oxford University Press.
Mills, K. (2017). 'Winning Humanitarian Interventions? Problematizing Victory and jus post bellum in International Action to Stop Mass Atrocities', in A. R. Hom, C. O'Driscoll, and K.

Mills (eds), *Moral Victories: The Ethics of Winning Wars*. Oxford: Oxford University Press. 156–175.

Mitchell, K. (2003). 'Monuments, Memorials, and the Politics of Memory', *Urban Geography*, 24(5): 442–459.

MOD (n.d.). *Development, Concepts and Doctrine Centre*. Available at: https://www.gov.uk/government/groups/development-concepts-and-doctrine-centre (Accessed: 12 January 2025).

MOD (2016). *Development, Concepts and Doctrine Centre: The MOD's independent think tank*. Ministry of Defence. Available at: https://assets.publishing.service.gov.uk/government/uploads/system/uploads/attachment_data/file/507544/20160217_DCDC_Pamphlet_Web__2_.pdf (Accessed: 18 February 2023).

Morgan-Owen, D., and Gould, A. (2022). 'The Politics of Future War: Civil-Military Relations and Military Doctrine in Britain', *European Journal of International Security*, 7(4): 551–571.

Moser, C. O. N. (2001), 'The Gendered Continuum of Violence and Conflict: An Operational Framework', in C. O. N. Moser and F. C. Clark (eds), *Victims, Perpetrators or Actors? Gender, Armed Conflict and Political Violence*. London and New York: Zed Books. 30–52.

Mycock, A. (2014). 'The First World War Centenary in the UK: "A Truly National Commemoration"?', *The Round Table*, 103(2): 153–163.

*Nasrallah victory speech, Bint Jbeil, Lebanon, 26 May 2000* (2019). Available at: https://www.youtube.com/watch?v=0YnH7qhENDQ (Accessed: 8 January 2023).

NATO (2017). 'AJP-01 Allied Joint Doctrine, Edition E Version 1'. NATO Standardization Office (NSO). Available at: https://www.gov.uk/government/publications/ajp-01-d-allied-joint-doctrine (Accessed: 12 March 2022).

Neumann, I. B. (1996). 'Self and Other in International Relations', *European Journal of International Relations*, 2(2): 139–174.

Neumann, I. B. (2018). 'Halting Time: Monuments to Alterity', *Millennium: Journal of International Studies*, 46(3): 331–351.

Nichols, R. (2010). 'Target Acquired: America's Army and the Video Games Industry', in N. Huntemann and M. T. Payne (eds), *Joystick Soldiers: The Politics of Play in Military Video Games*. New York: Routledge. 39–52.

Nieborg, D. B. (2010). 'Training Recruits and Conditioning Youth: The Soft Power of Military Games', in N. Huntemann and M. T. Payne (eds), *Joystick Soldiers: The Politics of Play in Military Video Games*. New York: Routledge. 53–66.

Nixon, R. (2011). *Slow Violence and the Environmentalism of the Poor*. Cambridge: Harvard University Press.

Nora, P. (1989). 'Between Memory and History: Les Lieux de Mémoire', *Representations*, 26: 7–24.

Nordin, A. H. M., and Öberg, D. (2015). 'Targeting the Ontology of War: From Clausewitz to Baudrillard', *Millennium: Journal of International Studies*, 43(2): 392–410.

Nordstrom, C. (2004). *Shadows of War: Violence, Power, and International Profiteering in the Twenty-First Century*. Oakland: University of California Press.

Nováková, L., and Šályová, R. (2019). 'Marking the Victory in Ancient Greece: Some Remarks on Classical Trophy Monuments', *ILIRIA International Review*, 9(1): 191–201.

Noyes, A. (1920). 'A Victory Dance', *The Saturday Evening Post*. June 19.

NSA Aap-06 (2013). 'NATO Glossary of Terms and Definitions (English and French)'. NATO Standardization Agency. Available at: https://www.jcs.mil/Portals/36/Documents/Doctrine/Other_Pubs/aap6.pdf (Accessed: 20 February 2023).

Obel Hansen, T. (2019). 'In Pursuit of Accountability During and After War', *Journal of Strategic Studies*, 42(7): 946–970.

O'Connor, R. G. (1969). 'Victory in Modern War', *Journal of Peace Research*, 6(4): 367–384.

O'Driscoll, C. (2015). 'At All Costs and in Spite of All Terror? The Victory of Just War', *Review of International Studies*, 41(4): 799–811.

O'Driscoll, C. (2019). 'Nobody Wins the Victory Taboo in Just War Theory', *Journal of Strategic Studies*, 42(7): 901–919.

O'Driscoll, C. (2020a). *Victory: The Triumph and Tragedy of Just War*. Oxford, New York: Oxford University Press.

O'Driscoll, C. (2020b). 'No Substitute for Victory? Why Just War Theorists Can't Win', *European Journal of International Relations*, 26(1): 187–208.

Oldenburg, D. (2002). 'Couch-Potato Commandos', *Washington Post*, 10 December. Available at: https://www.washingtonpost.com/archive/lifestyle/2002/12/10/couch-potato-commandos/bcd28f6d-a57b-46df-9fd1-2bd747085aa4/ (Accessed: 16 November 2022).

Osborne, B. S. (1998). 'Constructing Landscapes of Power: The George Etienne Cartier Monument, Montreal', *Journal of Historical Geography*, 24(4): 431–458.

OSCE (2007). 'Statement of the European Union on recent developments in Estonian-Russian relations'. Available at: https://www.osce.org/files/f/documents/1/8/24959.pdf (Accessed: 17 April 2021).

Palazzo, A. (2008). *From Moltke to Bin Laden: The relevance of doctrine in the contemporary military environment*. Study paper/Land Warfare Studies Centre, 315. Canberra: Land Warfare Studies Centre.

Palestrino, M. (2022a). 'Neglected Times: Laclau, Affect, and Temporality', *Journal of Political Ideologies*, 27(2): 226–245.

Palestrino, M. (2022b). 'Inking Wartime: Military Tattoos and the Temporalities of the War Experience', *International Political Sociology*, 16(3): 1–20.

Palestrino, M. (2025). '"Sow the Seeds of Victory": Wartime Gardening and the Ontology of Military Victory', *European Journal of International Relations*. OnlineFirst: https://doi.org/10.1177/13540661251322698.

Parashar, S. (2013). 'What Wars and "War Bodies" Know About International Relations', *Cambridge Review of International Affairs*, 26(4): 615–630.

Patterson, D. (2017). 'Victory and the Endings of Conflicts', in A. R. Hom, C. O'Driscoll, and K. Mills (eds), *Moral Victories: The Ethics of Winning Wars*. Oxford: Oxford University Press. 103–123.

Pearce, R. (2010). 'Tattoos—Life-Saving Art or Potential Health Hazard', *Journal of Military and Veterans' Health*, 18(4): 13–16.

Peic, G. (2014). 'Civilian Defense Forces, State Capacity, and Government Victory in Counterinsurgency Wars', *Studies in Conflict & Terrorism*, 37(2): 162–184.

Penney, J. (2010). '"No Better Way to 'Experience' World War II": Authenticity and Ideology in the Call of Duty and Medal of Honor Player Communities', in N. Huntemann and M. T. Payne (eds), *Joystick Soldiers: The Politics of Play in Military Video Games*. New York: Routledge. 191–205.

Pisani, D. (2017). 'From Italian Monuments to the Fallen of World War I to Fascist War Memorials', *RIHA Journal*. 30–44.

Pong, B. (2020). 'Wartime', *Modernism/Modernity Print Plus*, 5(2).

Puar, J. K. (2017). *The Right to Maim: Debility, Capacity, Disability*. Durham: Duke University Press.

Quinn, J., Joshi, M., and Melander, E. (2019). 'One Dyadic Peace Leads to Another? Conflict Systems, Terminations, and Net Reduction in Fighting Groups', *International Studies Quarterly*, 63(4): 863–875.

RAF (2009a). 'Air and Space Warfare AP 3002 (2nd Edition )'. Air Warfare Centre, Ministry of Defence.

RAF (2009b). 'British Air and Space Power Doctrine AP 3000 (4th Edition)'. Centre for Air Power Studies, Ministry of Defence.

Rao, R. (2019). 'One Time, Many Times', *Millennium: Journal of International Studies*, 47(2): 299–308.

Rao, R. (2020). *Out of Time: The Queer Politics of Postcoloniality*. Oxford, New York: Oxford University Press.

*Ready to Fight and Win | Who we Are | British Army* (2019). Available at: https://www.youtube.com/watch?v=yTaIKnRsaSU (Accessed: 12 March 2022).

Reardon, B. (1985). *Sexism and the War System*. New York: Teachers College Press.

Reiter, D., and Stam, A. C. (2003). 'Understanding Victory: Why Political Institutions Matter', *International Security*, 28(1): 168–179.

Robinson, N. (2012). 'Videogames, Persuasion and the War on Terror: Escaping or Embedding the Military—Entertainment Complex?', *Political Studies*, 60(3): 504–522.

Robinson, N. (2015). 'Have You Won the War on Terror? Military Videogames and the State of American Exceptionalism', *Millennium: Journal of International Studies*, 43(2): 450–470.

Robinson, N. (2016). 'Militarism and Opposition in the Living Room: The Case of Military Videogames', *Critical Studies on Security*, 4(3): 255–275.

Robinson, N. (2019). 'Military Videogames: More Than a Game', *The RUSI Journal*, 164(4): 10–21.

Rossi, P. J. (2015). 'Peacemaking and Victory: Lessons from Kant's Cosmopolitanism', *Philosophia*, 43(3): 747–757.

Rotte, R., and Schmidt, C. (2003). 'On the Production of Victory: Empirical Determinants of Battlefield Success in Modern War', *Defence and Peace Economics*, 14(3): 175–192.

The Royal Family (n.d.). Remembrance Day, The Royal Family. Available at: https://www.royal.uk/remembrance-day (Accessed: 16 November 2022).

Ryckman, K. C. (2020). 'Lasting Peace or Temporary Calm? Rebel Group Decapitation and Civil War Outcomes', *Conflict Management and Peace Science*, 37(2): 172–192.

Rynning, S., Schmitt, O., and Theussen, A. (eds) (2021). *War Time: Temporality and the Decline of Western Military Power*. Washington: Chatham House.

Sanders, C., and Vail, D. A. (2008). *Customizing the Body: The Art and Culture of Tattooing*. Revised and Expanded Edition. Philadelphia: Temple University Press.

Saunders, N. J. (2004). *Matters of Conflict*. London: Routledge.

Saunders, N. J. (2017). 'Trench Art: Objects and People in Conflict', in J. Bourke (ed.), *War and Art: A Visual History of Modern Conflict*. London: Reaktion Books. 209–215.

Scheipers, S. (2017). 'Carl von Clausewitz and Moral Victories', in A. R. Hom, C. O'Driscoll, and K. Mills (eds), *Moral Victories: The Ethics of Winning Wars*. Oxford: Oxford University Press. 34–52.

Schnapp, J. (2020). 'Small Victories ("BZ '18–'45")', in K. B. Jones and S. Pilat (eds), *The Routledge Companion to Italian Fascist Architecture*. Abingdon: Routledge. 533–545.

Scott, J. (2015). 'Objects and the Representation of War in Military Museums', *Museum and Society*, 13(4): 489–502.

Šisler, V. (2008). 'Digital Arabs: Representation in Video Games', *European Journal of Cultural Studies*, 11(2): 203–220.

Sjoberg, L. (2013). *Gendering Global Conflict: Toward a Feminist Theory of War*. New York: Columbia University Press.

Sloan, G. (2012). 'Military Doctrine, Command Philosophy and the Generation of Fighting Power: Genesis and Theory', *International Affairs*, 88(2): 243–263.

Smith, D.J. (2008). '"Woe from Stones": Commemoration, Identity Politics and Estonia's "War of Monuments"', *Journal of Baltic Studies*, 39(4): 419–430.

Smith, R. (2018). *11-11 Memories Retold Interview Talks Bringing WW1 to Players in a Different Way, TrueAchievements.* Available at: https://www.trueachievements.com/n34787/1111-memories-retold-interview-talks-bringing-ww1-to-players-in-a-different-way (Accessed: 16 November 2022).

Solomon, T. (2014). 'Time and Subjectivity in World Politics', *International Studies Quarterly*, 58(4): 671–681.

Solomon, T. (2015). *The Politics of Subjectivity in American Foreign Policy Discourses.* Ann Arbor: University of Michigan Press.

Solomon, T. (2019). 'Rhythm and Mobilization in International Relations', *International Studies Quarterly*, 63(4): 1001–1013.

Soragni, U. (1993). *Il Monumento alla vittoria di Bolzano: architettura e scultura per la città italiana, 1926–1938.* Vicenza: Neri Pozza Editore.

Stahl, R. (2006). 'Have You Played the War on Terror?', *Critical Studies in Media Communication*, 23(2): 112–130.

Stahl, R. (2010). *Militainment, Inc.: War, Media, and Popular Culture.* 1st Edition. New York: Routledge.

Steinacher, G. (2013). 'Fascist Legacies: The Controversy over Mussolini's Monuments in South Tyrol', *Faculty Publications, Department of History*, 144: 647–666.

Stevens, T. (2016). 'Governing the Time of the World', in A. R. Hom et al. (eds), *Time, Temporality and Global Politics.* Bristol: E-International Relations Publishing. 59–72.

Stevens, Q., Franck, K. A., and Fazakerley, R. (2018). 'Counter-Monuments: The Anti-Monumental and the Dialogic', *The Journal of Architecture*, 23(5): 718–739.

Stevenson, R. (2020). 'Time and Space Obliterated: Remembrance and Relativity in November1919', *Modernism/Modernity Print Plus*, 5(2).

Strachan, H. (2016). 'How Do Wars End? The Problem of Victory and Defeat', *The termination of wars in historical perspective.* National Institute for Defense Studies. 13-33.

Stroszeck, J. (2004). 'Greek Trophy Monuments', in S. Bouvrie (ed.), *Myth and Symbol II: Symbolic Phenomena in Ancient Greek Culture.* Bergen: Norwegian Institute at Athens. 303–331.

Subotic, J., and Steele, B. J. (2018). 'Moral Injury in International Relations', *Journal of Global Security Studies*, 3(4): 387–401.

Sullivan, P. L. (2007). 'War Aims and War Outcomes: Why Powerful States Lose Limited Wars', *Journal of Conflict Resolution*, 51(3): 496–524.

Sullivan, P. L. (2008). 'At What Price Victory? The Effects of Uncertainty on Military Intervention Duration and Outcome', *Conflict Management and Peace Science*, 25(1): 49–66.

Sullivan, P. L. (2012). *Who Wins?: Predicting Strategic Success and Failure in Armed Conflict.* New York: Oxford University Press.

Sullivan, P. L., and Karreth, J. (2015). 'The Conditional Impact of Military Intervention on Internal Armed Conflict Outcomes', *Conflict Management and Peace Science*, 32(3): 269–288.

Sylvester, C. (2009). *Art/museums: International Relations Where We Least Expect It.* Abingdon: Paradigm Publishers.

Sylvester, C. (ed.) (2011). *Experiencing War.* New York: Routledge.

Sylvester, C. (2012). 'War Experiences/War Practices/War Theory', *Millennium: Journal of International Studies*, 40(3): 480–503.

Sylvester, C. (2013a). 'Experiencing War: A Challenge for International Relations', *Cambridge Review of International Affairs*, 26(4): 669–674.

Sylvester, C. (2013b). *War as Experience: Contributions from International Relations and Feminist Analysis.* New York: Routledge.

Sylvester, C. (2018). 'Curating and Re-Curating the American War in Vietnam', *Security Dialogue*, 49(3): 151–164.

Sylvester, C. (2019). *Curating and Re-Curating the American Wars in Vietnam and Iraq*. New York: Oxford University Press.

Taylor, T., and Guyver, R. (eds) (2012). *History Wars and The Classroom: Global Perspectives*. Charlotte: Information Age Publishing.

*The Unwanted* (2014). 'War Ink—"11 for 11"'. Available at: https://vimeopro.com/murga/war-ink-project/page/2 (Accessed: 16 February 2023).

Tierney, D. (2015). *The Right Way to Lose a War: America in an Age of Unwinnable Conflicts*. New York: Little Brown & Company.

Tierney, D. (2017). 'The Ethics of Unwinnable War', in A. R. Hom, C. O'Driscoll, and K. Mills (eds), *Moral Victories: The Ethics of Winning Wars*. Oxford: Oxford University Press. 123–140.

Till, K. E. (2003). 'Places of Memory', in J. Agnew, K. Mitchell, and G. Toal (eds), *A Companion to Political Geography*. Hoboken: John Wiley & Sons. 289–301.

Tim (2018). *11–11 Memories Retold Developer Interview, Big Red Barrel*. Available at: http://www.bigredbarrel.com/blog/2018/11/11-11-memories-retold-developer-interview/ (Accessed: 16 November 2022).

Tribute Ink (2019a). *Paul Glazebrook | Tribute Ink Story | Royal British Legion, The Royal British Legion*. Available at: https://www.britishlegion.org.uk/get-involved/remembrance/tribute-ink/remembering-the-fallen/paul-glazebrook (Accessed: 15 January 2021).

Tribute Ink (2019b). *Craig Daniell | Tribute Ink Story | Royal British Legion, The Royal British Legion*. Available at: https://www.britishlegion.org.uk/get-involved/remembrance/tribute-ink/rethinking-remembrance/craig-daniell (Accessed: 15 January 2021).

Tribute Ink (2019c). *Tribute Ink | Remembrance | Royal British Legion, The Royal British Legion*. Available at: https://www.britishlegion.org.uk/get-involved/remembrance/tribute-ink (Accessed: 16 January 2021).

Tribute Ink (2019d). *Dani Cummings | Royal British Legion, The Royal British Legion*. Available at: https://www.britishlegion.org.uk/get-involved/remembrance/tribute-ink/marking-the-memories/dani-cummings (Accessed: 16 January 2021).

Tribute Ink (2019e). *Matt Tomlinson | Tribute Ink Story | Royal British Legion*. Available at: https://www.britishlegion.org.uk/get-involved/remembrance/tribute-ink/remembering-the-fallen/matt-tomlinson (Accessed: 16 January 2021).

*Tribute Ink—Craig Daniell* (2021). Available at: https://www.youtube.com/watch?v=46EQEufUw90 (Accessed: 16 February 2023).

*Tribute Ink—Danielle Cummings* (2021). Available at: https://www.youtube.com/watch?v=4dhu-1kcCFQ (Accessed: 16 February 2023).

Vale, L. J. (1999). 'Mediated Monuments and National Identity', *The Journal of Architecture*, 4(4): 391–408.

Van de Ven, A. (2020). 'Triumph (Re-)Imagined: Saddam's Monument to Victory', *International Journal of the Classical Tradition*, 27(1): 112–134.

Vaughan, M. (2007). 'Scarification in Africa: Re-Reading Colonial Evidence', *Cultural and Social History*, 4: 385–400.

War Ink (2014). *War Ink*. Available at: http://www.warink.org/ (Accessed: 15 January 2021).

*Watch The Army's New Recruitment Adverts for 2021 | Forces TV* (2021). Available at: https://www.youtube.com/watch?v=yTaIKnRsaSU (Accessed: 12 March 2022).

Wegner, N. (2021). 'Ritual, Rhythms, and the Discomforting Endurance of Militarism: Affective Methodologies and Ethico-Political Challenges', *Global Studies Quarterly*, (1): 1–10.

Welland, J. (2017). 'Violence and the Contemporary Soldiering Body', *Security Dialogue*, 48(6): 524–540.
Welland, J. (2018). 'Joy and War: Reading pleasure in Wartime Experiences', *Review of International Studies*, 44(3): 438–455.
Whitman, J. Q. (2012). *The Verdict of Battle: The Law of Victory and the Making of Modern War*. Cambridge: Harvard University Press.
Wibben, A. T. R. (2020). 'Everyday Security, Feminism, and the Continuum of Violence', *Journal of Global Security Studies*, 5(1): 115–121.
Wibben, A. T. R. (2011). *Feminist Security Studies: A Narrative Approach*. Abingdon: Routledge.
Wibben, A. T. R., Confortini, C. C., Roohi, S., Aharoni, S. B., Vastapuu, L., and Vaittinen, T. (2019). 'Collective Discussion: Piecing-up Feminist Peace Research', *International Political Sociology*, 13(1): 86–107.
Wibben, A. T. R., and Donahoe, A. E. (eds) (2020). *Teaching Peace and War: Pedagogy and Curricula*. Abingdon: Routledge.
Wilcox, L. B. (2015). *Bodies of Violence: Theorizing Embodied Subjects in International Relations*. Oxford, New York: Oxford University Press.
Wilkens, J., and Kessler, O. (2021). 'Concluding Chapter: Concepts at Work in Global IR', in P. Ish-Shalom (ed.), *Concepts at Work: On the Linguistic Infrastructure of World Politics*. Ann Arbor: University of Michigan Press. 203–221.
Wilkes, A. (2018). *Oxford AQA GCSE History: Conflict and Tension First World War 1894–1918*. Oxford: Oxford University Press.
Wilkinson, A. (2016). *AQA GCSE History: Power and the People*. UK Edition. London: Hodder Education.
Wilkinson, M. L. N. (2014). 'The Concept of the Absent Curriculum: The Case of the Muslim Contribution and the English National Curriculum for History', *Journal of Curriculum Studies*, 46(4): 419–440.
Willasey-Wilsey, T. (2021). 'Twelve Dilemmas Behind the UK's Afghan Defeat', *RUSI*, 2 August. Available at: https://www.rusi.orghttps://www.rusi.org (Accessed: 25 January 2023).
Willis, M., and Thomas, J. (2016). *Oxford AQA History A Level and AS Component 2: Wars and Welfare: Britain in Transition 1906–1957*. Oxford: Oxford University Press.
Winter, C., and Mills, C. (2020). 'The Psy-Security-Curriculum ensemble: British Values curriculum policy in English schools', *Journal of Education Policy*, 35(1): 46–67.
Wool, Z. H. (2015). *After War: The Weight of Life at Walter Reed*. Durham: Duke University Press.
Yadav, P., and Horn, D. M. (2021). 'Continuums of Violence: Feminist Peace Research and Gender-Based Violence', in T. Väyrynen, S. Parashar, É. Féron, and C. C. Confortini (eds), *Routledge Handbook of Feminist Peace Research*. 1st Edition. Abingdon: Routledge. 105–114.
Young, J. E. (1992). 'The Counter-Monument: Memory Against Itself in Germany Today', *Critical Inquiry*, 18(2): 267–296.
Zaidi, B. M. I. (2009). *The Conduct of War and the Notion of Victory: A Theory and Definition of Victory*. PhD Thesis, Cranfield University.
Zisser, E. (2010). 'Israelis Confront the Second Lebanon War', *Bustan: The Middle East Book Review*, 1(1): 29–43.

# Index

*For the benefit of digital users, indexed terms that span two pages (e.g., 52–53) may, on occasion, appear on only one of those pages.*

The manufacturer's authorised representative in the EU for product safety is Oxford University Press España S.A. of el Parque Empresarial San Fernando de Henares, Avenida de Castilla, 2 – 28830 Madrid (www.oup.es/en or product.safety@oup.com). OUP España S.A. also acts as importer into Spain of products made by the manufacturer.

www.ingramcontent.com/pod-product-compliance
Lightning Source LLC
Chambersburg PA
CBHW070849080326
40795CB00045B/428/J
* 9 7 8 0 1 9 8 9 5 5 7 8 8 *